Estimating & Bidding
for Builders & Remodelers

By Richard J. Langedyk, Senior Instructor
Construction Estimating Institute of America

Craftsman Book Company
6058 Corte del Cedro / P.O. Box 6500 / Carlsbad, CA 92018

[Acknowledgemen]ts

The author and publisher wish to express their appreciation to the following for furnishing materials used in the preparation of various portions of this book:

Construction Specifications Institute, 601 Madison Street, Alexandria, Virginia 22314

Merillat Industries, Inc. P.O. Box 1946, Adrian, Michigan 49221

The Library of Congress has cataloged the first edition of this book as follows:

Library of Congress Cataloging-in-Publication Data

Langedyk, Richard J.
Estimating & bidding for builders & remodelers / by Richard J. Langedyk.
p. cm.
Includes index.
ISBN 0-934041-66-0
1. Building—Estimates. 2. Building—Repair and reconstruction--Estimates. I. Title. II. Title: Estimating and bidding for builders and remodelers.
TH435.L25 1992
692'.5—dc20 92-23271
CIP

Second edition ISBN 1-57218-010-2
Third edition ISBN 1-57218-027-7

Contents

Chapter 1
Get Started Right 5

Sink or Swim in Construction Contracting. .6
The Detailed Cost Estimate 7
Estimating as a Career 8
Typical Estimator Profile 10
Shortcomings of an Estimating Career . . .12
Rewards 12

Chapter 2
Planning Your Estimate 17

It All Starts With a Plan 17
Custom-made Markup 18
The Long-term Plan 21

Chapter 3
Getting Ready to Estimate 29

It All Starts With Plans 29
The Plan Package 32
Examine the Site .36
The Cost Data File 36
Doing the Take-off 38
Compiling an Estimate 39
Checking Estimates 47

Chapter 4
The Estimating Process 51

Detailed Estimating Steps 51
A Sample Take-off 63
The Detailed Estimate Advantage 88

Chapter 5
Estimating Repair & Remodeling Work 89

Limiting the Scope of Your Bid 90
Protecting Yourself 91
A Sample Remodeling Estimate 93
The Estimating Process 96
Putting it All Together 111
Checking the Estimate 117
Preparing the Bid 117
Closing the Sale .120

Chapter 6
Estimating Commercial Work 125

The Commercial Estimate 126
Follow a Plan 128
Visit the Site 129
Start the Take-off 129
The Take-off Form 131
How to Take off Quantities 132
Applying Unit Costs 146

Chapter 7
Pricing With a Computer 153

Using CD Estimator 154
Installing CD Estimator 156
Using National Estimator 157
Opening Other Costbooks 167
Contractor's Home Page on the Internet .168
Contractor's BBS 168
Installing Excalibur 169
First Time Log Ons 172
Downloading Revised Cost Files 173
On-line Assistance 175

Chapter 8
Cost Recording 177

Why Keep Cost Records? 178
Essentials of a Cost System 179
Cost Records Check Current Jobs 182
Classifying Labor Costs 182
Using Cost Data 187

Chapter 9
Planning Overhead 191

Start With Last Year 192
The Annual Budget 196
My Budget for AEI Builders 200
The Preliminary Budget Bottom Line . . . 208
Failing to Plan Is Planning to Fail 213

Chapter 10
Estimating Overhead & Profit 215

Your Normal Markup 216
The Right Markup for Your Estimates . . 218
But Will We Make More Money? 222
Develop a Profit Curve 222
Your Own Best Profit Range 227

Chapter 11
Smart Bidding 229

Adjustment for Risk 229
Asset Utilization Adjustments 240
Identify Under-utilization 242
Project Adjustments 248
What Is the Net Effect? 251

Chapter 12
Pricing Strategies 253

Learn About Your Competition 255
Graph Your Competition 260
Let's Wrap This Up 264

Index 265

Chapter 1
Get Started Right

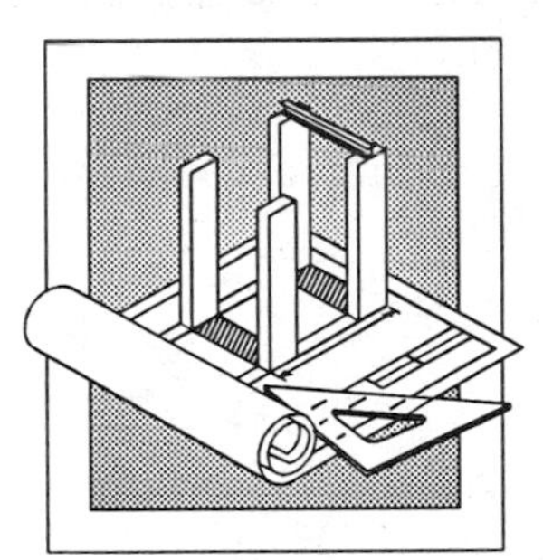

It's been said that construction estimating is more an art than a science. Truly, it's both. Like scientists, estimators collect what they hope is accurate information: precise take-off quantities, exact price quotes and carefully-documented costs of completed jobs. But like artists, estimators rely on experience, intuition, judgment, and sometimes, guesses.

Why can't construction estimating be more science and less guesswork? If you've done much estimating, the answer is probably clear. Estimating will never be a science. Every construction project is unique. No single cost can be accurate for all jobs and all bidders. Estimates have to be custom-made for the job, the time, the place and the crew that's going to do the work.

And that's what this book is about: The fine art of accurate construction cost estimating.

The fine art of cost estimating

This guide is written for the owners, estimators and project managers in small and medium-sized construction companies. I'll take a scientific approach to estimating when that's possible. I'll show you how to eliminate a lot of the uncertainty in bidding and tip the scales in your favor when that's an option. When it's not, I'll suggest easy ways to limit your risk and reduce your exposure to a major loss.

My goal throughout this book is to help you make a good living as a construction estimator. It's possible. In the U.S., hundreds of thousands of construction cost estimators do just that. If you aren't making a good living now, I expect this book will help you.

Construction Is the Largest Industry

The Small Business Administration has estimated that there are over 500,000 "visible" construction businesses with one or more employees. In addition, there are about 250,000 independent operators in construction. Contract construction is the largest industry in the United States by number of people employed and accounts for between 10 and 11 percent of the gross national product. Most of this work and most of the profits go to the larger firms with 25 or more employees. Yet there are a large number of smaller firms eager to expand and take on additional business. Since competition in construction is largely price competition, the level of profit for everyone tends to be low. Since many construction companies are thinly capitalized, a low profit margin can result in a disastrous loss if the job doesn't go as estimated. In spite of the low overall profit margins, the growth potential is great because along with high risk goes the potential of high gain.

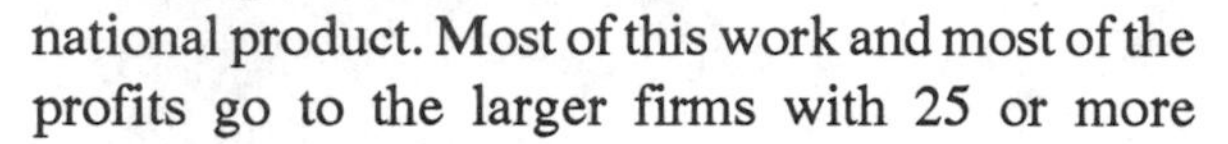

First, I'll explain the basics, how to use the construction documents (the contract and specifications) and create a good material take-off from the plans. That's the subject of the first seven chapters. Beginning in Chapter 8, I'll tell you about more advanced construction cost estimating and bidding techniques.

I'm going to assume that you already know how to read plans. If you haven't developed good plan-reading skills, several books listed at the end of this manual can provide all the practical information and details you need.

The information I present here isn't restricted to any particular construction trade. This book will be as useful to remodeling estimators as it is to general construction estimators. No matter what size your jobs (from $5,000 to $5,000,000), no matter what type of work (from foundations to roofing and everything in between), this manual will help you prepare more accurate construction cost estimates.

Sink or Swim in Construction Contracting

I believe there are five key areas that make or break most construction companies. None involve construction skills like driving nails or reading plans. Every construction company has (or can hire) skilled workers and supervisors to do the actual building.

My five key skill areas are almost always the responsibility of the owner (or senior estimator) in a construction company. All five are either a part of the estimating process or influence cost estimates. All affect both profitability and survival of the company.

Five basic skills everyone should master

Here's my list:

- Finding and bidding the type of work that's appropriate for your company
- Making accurate cost estimates
- Using pricing strategies that reduce your risk of loss
- Anticipating the bids of your competition
- Tracking overhead so you know the actual cost of doing business

As you've probably guessed already, the focus of this book is to explain what you should be doing in each of these areas. Your company can probably survive (for a while) without doing anything I recommend here. But follow my advice and you're almost guaranteed to do better than you're doing now. That's my promise. And I hope it's your reason for reading this book.

The Detailed Cost Estimate

Let me begin by making one very important point. Throughout this book I'll be talking about estimates. By that I mean a detailed labor and material cost estimate. That's a list of every work item in the job, with prices for all labor, material and equipment needed to complete that unit of work. There are many other ways to estimate costs. We'll discuss some. But the most accurate estimate is always the detailed labor and material unit cost estimate.

Examples of detailed cost estimates

What's a unit? Here are some examples:

- Cubic yards of concrete, purchased and placed
- Tons of reinforcing steel, purchased and placed
- Square feet of drywall, hung, taped and textured
- Square yards of floor covering, including installation

Every general condition (project overhead) item can be estimated by the unit cost. For example, the porta-john you'll need on site costs $50 a month (including delivery, weekly service and pickup) and you need it for three months. The cost of that work unit is $150 (3 months times $50 per month).

If you're bidding a 100,000 square foot building, there may be 1,000 unit cost items in the estimate. The only way to compile an accurate estimate is to find each of those 1,000 cost items. How do you do it?

There's only one answer to that question: by studying the plans, specs and construction documents. Every time you find a cost item, write it down on your take-off sheet. If there are 1,000 items, you'll fill about 1,000 lines on your estimating pad. When your take-off is complete, write in an estimated labor, material and equipment cost on each line. That's detailed cost estimating. It takes time. It's not easy. And it's expensive. But it's the only way I know of to prepare *accurate* construction cost estimates.

There's no such thing as a quick, simple, easy way to estimate construction costs. It takes time, effort and attention to detail. But there are easier, quicker, more consistent ways to estimate costs. I'll cover them, including estimating with a computer. As you may have already discovered, there's a CD-ROM disk in an envelope bound inside the back cover of this book. After we've covered manual (pencil and paper) estimating procedures, I'll give you some practice in writing estimates with a computer.

"Only the best estimators can scope out the last 10 percent . . ."

There's a big difference between an acceptable estimate and a truly first class estimate. Most estimators can get within 90 percent of the actual construction cost most of the time. All that takes is attention to detail and hard work. Only the best estimators can scope out the last 10 percent — and maybe spot a problem or major omission that could turn a potential money-maker into a big loss.

Later, in Chapter 4, I'll ask you to examine your own estimating system and compare it to the detailed system I recommend. I'll also suggest some shortcuts that can increase productivity and streamline your estimating procedure, no matter what method you use.

Estimating as a Career

I'm not going to get into the meat of construction estimating without providing a little pep talk about estimating as a career. Like any good teacher, I want to be sure you have the motivation to absorb what I'm going to explain.

In my opinion, cost estimating is one of the most neglected career opportunities in the U.S. today. I've never seen any published figures, but my guess is that there are about one million estimating jobs in the United States. Why so many? Because costs for every home, apartment, office, store and factory building are estimated and re-estimated many times before that building is finally demolished.

Even then, the cost of demolition has to be estimated. Nothing gets designed or built or remodeled or insured or taxed or torn down without some estimate of the cost.

You've Got Job Security

Of course, most construction cost estimating happens before a building goes up. You can be sure there's at least one estimate prepared by every bidder. And even a modest remodeling project may have dozens of bidding contractors and subcontractors — carpentry bids, plumbing bids, electrical bids, concrete bids, and so on. Every construction project begins with a blizzard of estimating paperwork. If you're in construction and not into estimating, maybe you should keep reading. There's an opportunity here.

I feel that estimating is good work. It pays well, and it's a job that carries responsibility and earns respect. Maybe that's why estimators have more job security than most people in construction. The chief estimator is going to be the last person fired in any construction office. After all, a construction company that's stopped estimating is out of business. It's the estimator who brings in new work. It's the estimator who understands the company finances, sets profit margins and controls volume. It's the estimator who makes the difference between financial success and failure in most construction companies.

So, what does it take to be a construction estimator?

General and Special Qualifications

First of all, understand this. Estimating is an accidental profession. Our public schools don't teach it. I've never had a little kid tell me that he (or she) wanted to grow up to be a construction estimator. Yet many will. And even more should. Why? Because there are no barriers to entry into this profession. States don't license construction cost estimators. You don't need a graduate degree. You don't have to pay a fee to some government agency to call yourself a construction estimator. Anyone can be an estimator.

Working your way up the career ladder

Many estimators started out as construction tradesmen and worked their way up to a desk job. Others wanted to be architects and never finished school. So rather than drawing plans, they settled into reading plans and doing take-offs. No matter how you get there, you have to qualify yourself to estimate construction costs and make a good living at it.

So, what does it take to be a good construction cost estimator? It's obvious that estimators have to read and understand construction drawings. Blueprints are the language construction professionals use to communicate.

Plan reading is essential. Basic math skills are important, too. If you can't read plans and add a column of figures, you're already a step behind as an apprentice estimator.

". . . Good estimators have two seemingly contradictory skills. They're both generalists and specialists."

Of course, nearly anyone can learn to read plans and use a calculator. But it takes more to become a skilled professional estimator. I've found that most good estimators have two seemingly contradictory skills. They're both generalists and specialists. They see every one of the trees without losing sight of the forest.

As generalists, they have a good grasp of the big picture. The best estimators seem to understand intuitively how all the parts, all the trades, will come together to create the whole.

Second, they're specialists, focusing intently on the details. It's common to work for hours on a single complex drawing, identifying every work item and the labor required to complete what the plans show. The best estimators are very good at that kind of detail work.

But details alone don't make a project. Without a broad view of how the parts come together to create the whole, something will be left out. Some cost is going to be omitted. When that happens, the estimated cost isn't going to match the actual cost. That's expensive, at best, and may be a financial disaster. Either way, it's bad estimating.

Typical Estimator Profile

I've taught and talked with hundreds of construction estimators from construction companies in all parts of the country. My guess is that about two-thirds of these estimators are owners of small to medium-size construction companies. They manage the company, run the crews, bring in the work and prepare the estimates. Most of these estimators began as apprentice tradespeople. They progressed quickly to journeyman status, then supervisor, usually because they learned faster, worked harder and had more ambition than others in their position.

Sooner or later, these bright, ambitious, energetic tradespeople got tired of working on someone else's payroll. They were anxious to venture out on their own, bidding for their own jobs. Unfortunately, most do this with little or no formal training in construction cost estimating. Instead, they rely on trial and error, getting experience the hard (and expensive) way. Eventually they either learn the skills they need to survive, or go back to working for wages.

Most estimators started as apprentice carpenters. That's because most entry-level jobs in the construction industry in this country are in erecting wood-frame buildings. I believe that carpentry is good basic training for estimators. It's where I started. And I don't regret the years I spent on a carpentry crew. It exposed me very quickly to the entire construction process, from setting foundation forms to framing the roof.

Twenty years ago, probably less than 25 percent of all estimators had more than a high school diploma. I sense that's changing. We're a better-educated nation today than we were in the 1970s. More and more estimators have college degrees or college-level training in construction technology, engineering or architecture. That's good. In my opinion, better-educated, better-trained estimators make better estimators.

Position in Company

Most construction estimators don't estimate full time. As I said, most estimators also own and operate a construction contracting or subcontracting business. They don't trust anyone but themselves to make important decisions about costs and bid prices. Even if they could trust someone else to estimate costs, they can't afford to hire a professional construction estimator. So they put on an estimator's hat when it's estimating that has to be done.

Classifications of estimators

Larger construction contracting companies (more than $2,000,000 in annual revenue) usually have a staff estimator. These employee-estimators usually fall into one of three classifications. The first is beginners, or *junior estimators*. They do the measurements and take-offs. They study the plans, determine quantities, and apply material prices.

As they become more experienced, junior estimators are promoted to *journeyman estimator* rank, where they may assume job management responsibilities. In many companies, when an estimator prepares a winning bid, the estimator becomes the project manager, overseeing the work until it's completed.

The third class of estimator usually answers to the title of *chief estimator*. He (or she) is the senior person in the estimating department and probably the number two person in the company, reporting directly to the company owner.

Shortcomings of an Estimating Career

Most professions have disadvantages. In estimating, it's the constant disappointments and intense competition. I'll explain.

The estimator's job is to make an accurate estimate with prices low enough to win the job, but high enough to earn a reasonable profit. Most of the time, that doesn't happen. In fact, an estimator who is successful one time in four is the happy exception, not the rule. This is one career where there are more defeats than victories. No lawyer, doctor, teacher or professional baseball player could get by with a construction estimator's batting average. If you make good money on as few as 10 percent of the jobs you bid, you belong in the Estimating Hall of Fame.

Learn to defend your actions

Even on the jobs you win, you may have to defend yourself. Someone is going to ask, "Why was our bid so much lower than the next lowest bidder? You left too much money on the table. We could have made thousands more." When your bid is the lowest by far, your first thought is probably, "What did I miss?" That's going to be a major issue if you have to apply to a bonding company for a performance bond. Try to justify a price that's low by $100,000 on a $1,000,000 job!

Rewards

For me, the benefits of this profession far outweigh the burdens. First, I like building. Everything I bid and build will be around long after I'm gone. That gives me a sense of accomplishment, a feeling of pride. My children and grandchildren will remember me for the monuments I've left behind. Of course, I'm not the only one who can claim responsibility for these buildings. But my role was important, probably as important as anyone who worked on the job.

". . .The benefits of this profession far outweigh the burdens."

And I like being paid well for what I do. I'm a decision-maker. I evaluate risks and rewards. I try to make good choices. If I choose wisely, I'm entitled to a premium for making and saving money. If I don't get that premium, I can take my resumé to the competitor down the street, along with everything I've learned about making money for a former employer. That gives me leverage that a tradesman or even a supervisor doesn't have. In good times or bad, in boom or bust, I like that advantage. I'm not going to abuse the privilege. But I have it. And I'm going to use it to get what I feel is fair treatment.

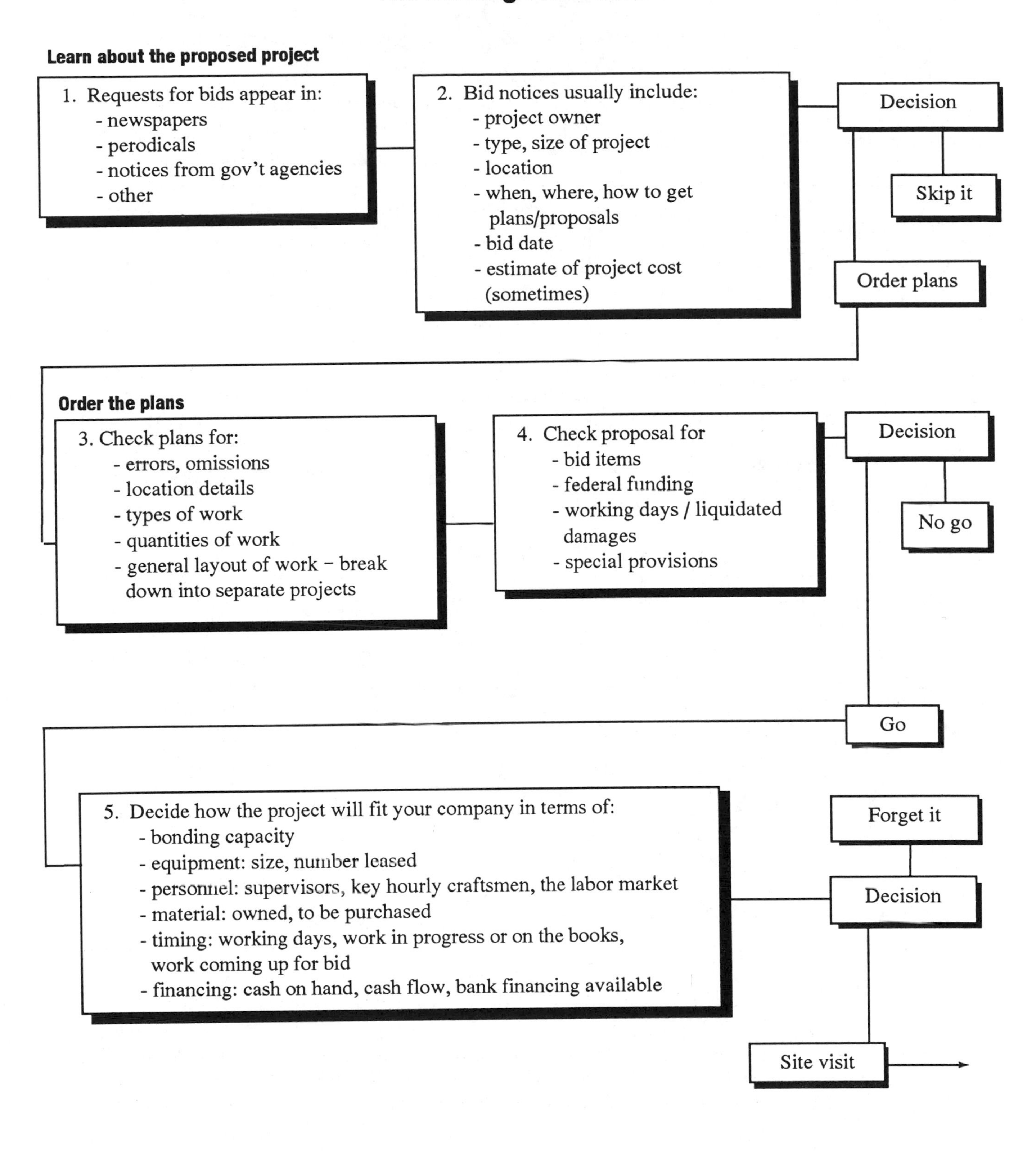

Figure 1-1 The bidding procedure

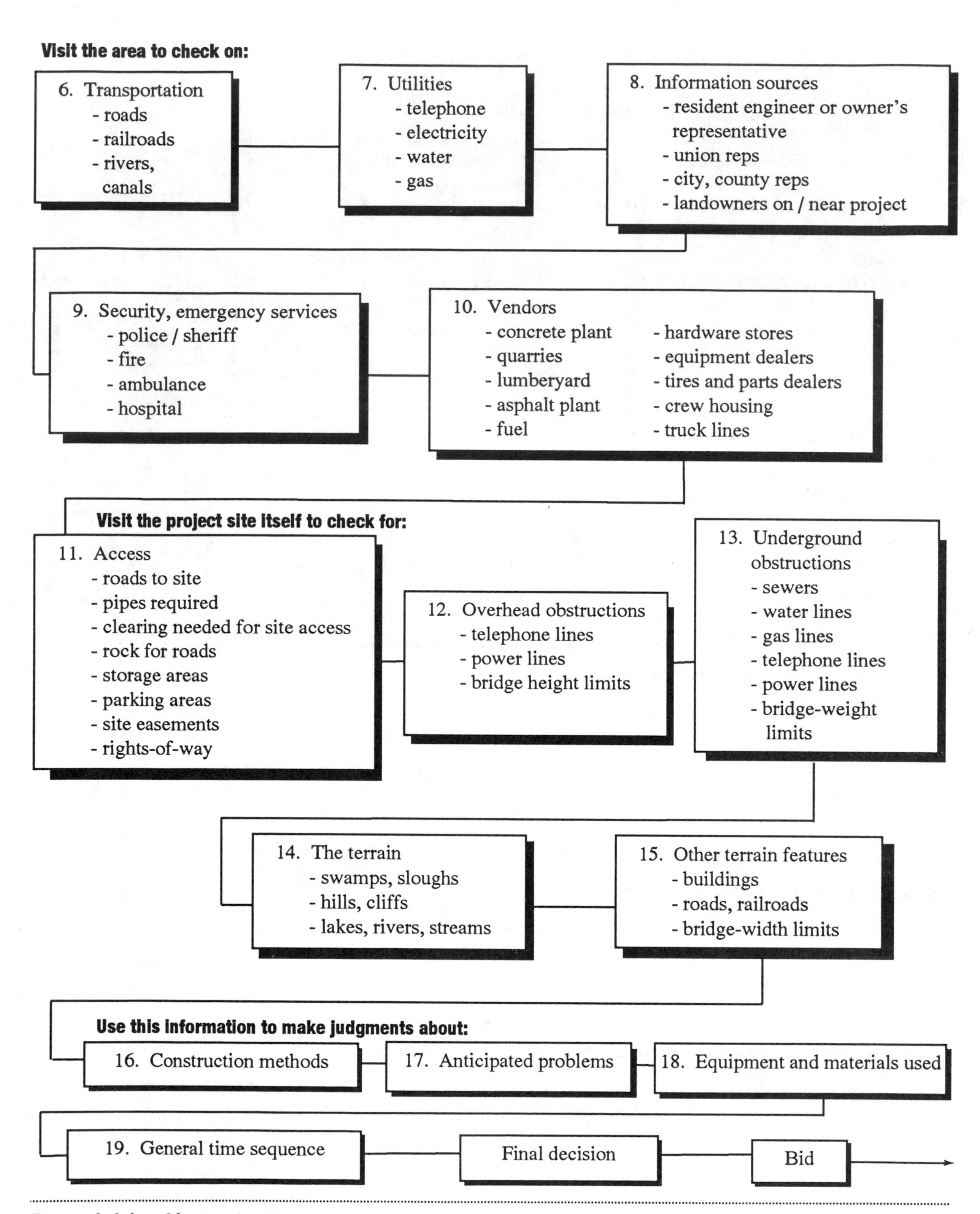

Figure 1-1 (cont.) The bidding procedure

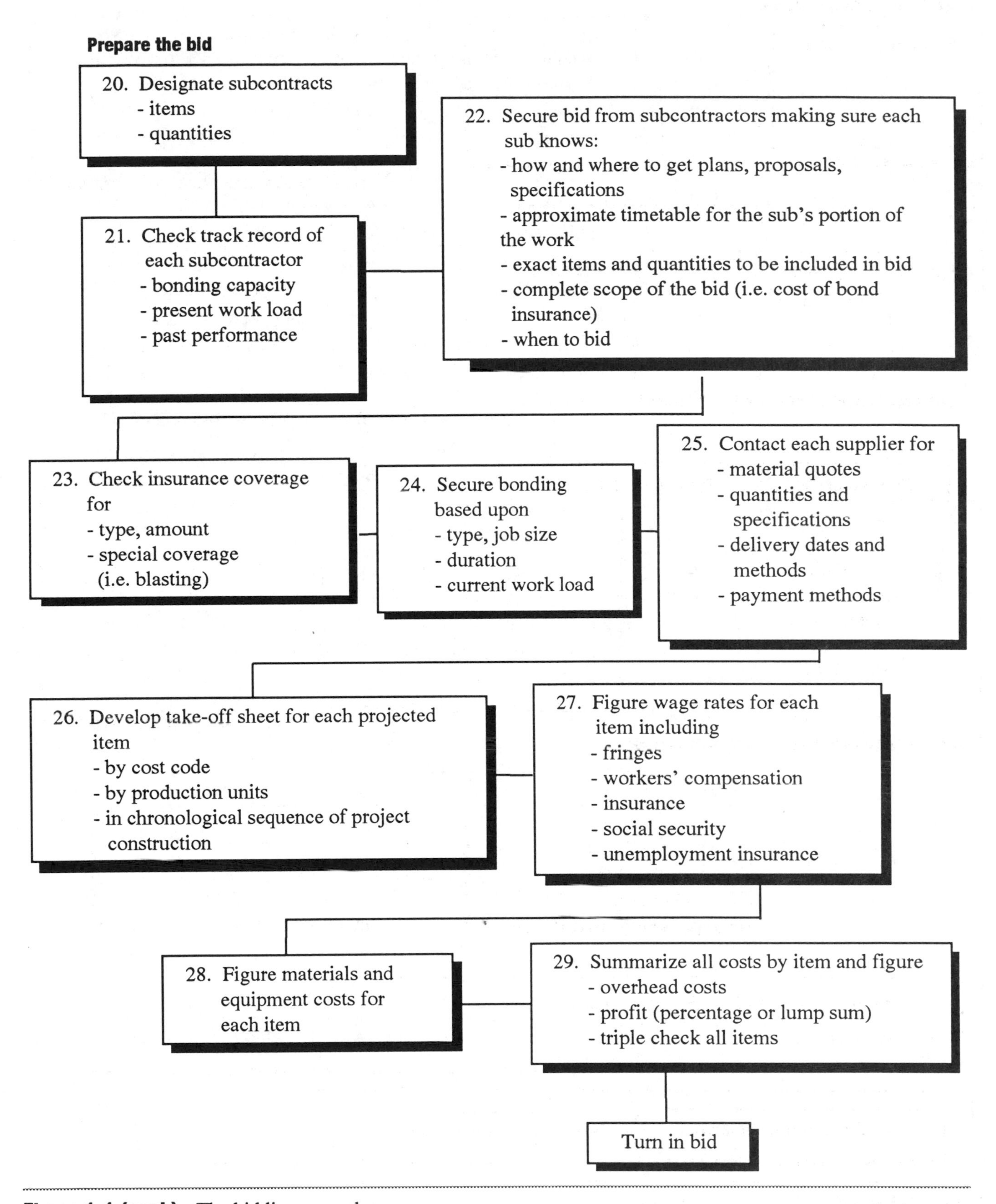

Figure 1-1 (cont.) The bidding procedure

Decision Making

As I said, estimators are decision makers. If you haven't thought about that before, think about it now. Every estimate, every project, presents a complex series of options. Make good decisions and you make good money. Make too many bad decisions and you look for a different job.

Think of every project as a decision chain. Every decision on that chain is linked to a decision made previously and to a decision yet to be made. Figure 1-1 shows the decision chain I'm talking about.

Our starting point is always the most basic decision of all. What jobs are we going to bid? Of all the work available, what jobs do we want? Pick the wrong jobs and you squander company assets (like your time, company working capital, and management talent). What you want is a good match between what you can handle well and what the project requires. Of course, there are other considerations. What's the competition? Do we need more work? Do we have the bonding capacity? What's the risk associated with this project? Decisions further down the chain in Figure 1-1 follow logically after we've made a tentative decision to bid a job.

Review the remainder of Figure 1-1 and you'll begin to get an appreciation of the size of our task. I'm going to cover all these decisions in the chapters that follow. Don't get discouraged by the size and complexity of the job. I'm going to explain it all one step at a time in simple language anyone can understand.

Chapter 2 begins with the most basic decision of all, "What work am I going to estimate?" Chapter 2 will also help you set goals for profit margin and volume for your company. These standards will be very important when we get into bidding strategy in Chapter 11.

Chapter 2
Planning Your Estimate

Before we unroll a set of plans and roll up our sleeves, I want to cover an important topic that's easy to miss: "What jobs should I bid?" That's a basic business planning question. You're really asking, "What business do I want to be in?"

In the last chapter I said that good estimators can see the forest, and every one of the trees too. The planning I'm talking about here is like the forest. It's the big picture, the part that's obvious if you think about it for a minute. Unfortunately, some of us estimators never do. And that's a mistake I hope you'll avoid.

My primary focus here will be to keep you out of a declining business, in a declining community and in a declining economy. First, ask yourself, "Is there money to be made bidding this type of work?" If not, I suggest you move on to something else.

Second, once you're estimating costs for the right work in the right community with the right economic conditions, set some goals and standards. These goals (volume levels and profit margins) become very important later when we start adding profit to our estimates. I'm going to suggest that you set a "normal" markup that you'll use on all jobs, unless there's a good reason why a normal markup shouldn't apply.

It All Starts With a Plan

Questions to ask yourself

Start the planning process by asking yourself some very basic questions.

- What kind of work do we want to handle?
- What's our annual gross income?

- What's our profit in percent of gross?
- How would volume change if we increased or decreased the profit margin?
- Could we increase volume and profits if we changed the focus of our business?
- Where do we want this company to be in five years?
- What are we doing to get there?

These are important questions for anyone running a construction company. They're also important questions for cost estimators. And, they're all business planning questions.

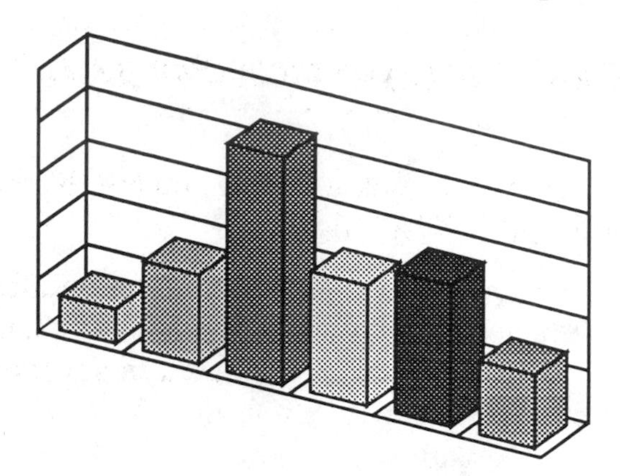

You might ask, "Why bother with business planning? Isn't that for huge companies like General Motors and IBM?" Sure, big companies plan. They have to. But you should too, no matter how big or small your company may be. Every construction company owner and estimator needs a hard-headed, no-nonsense business plan. That plan is a key part of every estimate you make. It's both a goal and a strategy for getting to that goal. Without it, you're going to find all sorts of excuses for making the same mistakes you've made in the past. With a plan, at least you know what you should be doing, even if you don't do it every time.

Later, in the chapter on advanced estimating, I'll refer repeatedly to your business plan. For now, let me sketch out just the essentials of your business plan.

Part of your plan should include an answer to the question, "How much overhead and profit should I include in my bids?" Obviously, that's an estimating question.

You probably have a profit goal already. Many contractors routinely include a 10 or 15 percent profit in their estimates. Others include whatever profit they think they can get away with. Some use a higher profit percentage on work that involves more risk or more uncertainty. I'm going to suggest a different approach.

Custom-made Markup

Is the profit percentage you use now the right one for your company? It's probably about right if you're still in business. If it were way too high or way too low, you probably would have gone out of business last week or last year. If it's a little too low, you're working for nothing. That gets discouraging after a while. It also drives contractors out of the business in a year or two — usually because they've run out of working capital.

Make your markup reasonable

If your profit percentage is too high, you won't get enough work. The market just won't support the price you're asking.

So if you've been in business for several years, the markup you're asking must be reasonable. But that's not to say it's perfect. If you charged a little less, maybe volume would increase, leaving more profit at the end of the year. Maybe you could charge just a little more, even 2 percent more, and still get all the business your company can handle.

". . . Pick projects that help you meet plan goals."

Exercises and examples later in this book will help you test the markup you're using and make adjustments if it isn't right.

Once you've got a business plan and have set a normal (or business plan) markup, the next step is to pick projects that help you meet plan goals. The objective is to stay busy on jobs likely to earn a reasonable return on your effort.

Varying Your Business Plan Price

My experience is that most construction contractors, including those who have a formal, written business plan, actually charge their normal business plan price less than half the time. Why not the rest of the time? Here are some reasons:

Be ready to adjust your bid

1) Business is slow. You need work badly. You have an opportunity to bid a job that could keep your company busy for several months. You know you have little chance of getting it at your normal markup. So you decide to knock off $5,000 and improve your chance of getting the job.

2) There's risk in the project. You're not comfortable with your estimate. How do you make up for that financial risk? You increase your normal bid price. That way, if you get the job, you're protected.

3) You know the competition on this bid. You know how they usually price their work. Maybe you'll cut a little off the bid to stay competitive. Or, you might know that some of the competition will be submitting courtesy bids, figures they feel are too high to be accepted. So you can afford to pad your figures a little. You'll be competitive even at higher markups.

These are only three of the many reasons for varying from your normal markup. We'll look into each of these situations in Chapter 10.

I agree that there are valid reasons to vary from your normal markup. But it's easy to get into trouble here — not with the decision to raise or lower the bid, but with the amount of the adjustment.

Where did the $5,000 come from in paragraph 1) above? Typically, it comes right out of the owner's or estimator's pocket. There's no justification for that number, other than that it's perceived as necessary to make the bid competitive. That may be a good enough reason, but maybe not the best one. I'll show you how to decide *exactly* how much to adjust your price.

And I'll take it a step further. Sometimes you may have to decide if you should take a job at no profit, or even at a loss, just to keep your crews on.

Some of these concepts and techniques may be new to you. Most people take a personal approach to decisions like discounts. But I'll show you how to use real comparisons to reach these pricing decisions, and to base your decisions on how they will affect your profit pattern over the next month, quarter or year.

Market Conditions and Pricing

Every contractor's business plan has to account for one important fact: There's a distinct economic cycle in the construction industry. Generally it lasts about eight years from peak to peak.

"During a recession, structure your business so it's profitable . . ."

If you've been in construction more than 20 years, you certainly remember 1974, 1982 and 1991. Those were bad years in the construction industry. There wasn't much work available and competition was intense. More than a few construction companies went belly up. But just as surely as there are busts, there are booms. It's a regular cycle and has been fairly predictable since World War II ended the Great Depression. Expect it to continue. Include the economic cycle in your business planning.

Construction is usually the first industry to be affected by an economic slump. Builders are forced to slice markups to less than they consider normal. You can't fight that. Instead, expect it and plan accordingly.

During a recession, structure your business so it's profitable at lower margins. The most recent construction recession of 1991 drove prices and markups down — about 10 percent for many contractors. Most had to trim overhead to stay competitive. Nearly all cut unnecessary expenses to the bone.

In the mid-1980s many contractors sent their staff estimators to estimating conventions in company-owned aircraft. That didn't happen in 1991. Companies had to cut back their overhead.

There was a time in the mid-1980s when a small developer or spec builder could count on selling mid-range single-family homes well before they were finished. But in 1991 it was common to see spec-built homes and commercial buildings sitting vacant, even years after completion. Contractors who had little competition during the 1980s suddenly had to bid against 15 or 20 other builders on every significant project. Many companies that prospered in the 1980s disappeared in the early 1990s. To get any work, builders had to slice overhead and provide special services to their clients.

In slow times, contractors everywhere have to structure their companies so they're profitable with less volume and at lower margins. Some are forced to switch markets. That usually means moving to a new area where work is available.

Here's one thing you *can't* do in a slowdown: Sit there and wait for the market to improve. That's usually a prescription for disaster.

The Long-term Plan

A good construction company business plan will control company growth, improve market position, increase profit and use staff and equipment more efficiently.

I'm recommending that you create a long-term business plan and revise it periodically — at least once a year. Make revisions based on your company's performance during the past year and your predictions of what the construction market promises for the future.

"Even the most competent manager, working with adequate resources and high motivation, won't succeed in construction . . . without a set of goals."

Essentials for a long-term plan

The following elements are essential for developing a successful long-term business plan:

- Broad employee participation
- Realistic market projections
- Recognition of required resources
- Strategic competitor awareness
- Effective implementation

Even the most competent manager, working with adequate resources and high motivation, won't succeed in construction (or any other business) without a set of goals. Think of a business plan as your road map to success. If it's realistic and you follow it, it's a major asset to your company. If it's based on miscalculations, it's worse than not having a plan at all.

A business planning book written specifically for contractors is *Contractor's Growth & Profit Guide* by Michael C. Thomsett. The order form bound into the back of this book tells you how to order a copy. The book explains exactly what should go into your business plan. I'm not going to repeat here what's explained very completely in the Thomsett book. In fact, I'm not even going to insist that your business plan has to be written down on a piece of paper. (But if you're applying for a loan, your lender definitely will want a written plan.) For our purposes, a plan you carry around in your head may be almost as good. In some ways, it may be better. The important thing is that you have a plan, have set goals, and are working toward those goals every day.

Design your business plan to cover the next five to seven years. You can't realistically plan beyond that time because you don't have enough information. You can't predict the construction market that far into the future. Five to seven years is enough.

Questions to ask yourself

How do you begin? Start with information about where you are today:

- What size is your business?
- What is your product specialty?
- What is your market area?
- How aggressive is your competition?
- What was your gross revenue last year?
- What was your markup on that volume of work?

Next, take a guess at what your market will be like in five years if you keep doing exactly what you're doing now. Be a little pessimistic. Assume your construction market will decline (even if temporarily) in the next five years. Adjust your long-term business plan accordingly.

What You Want vs. What You Can Have

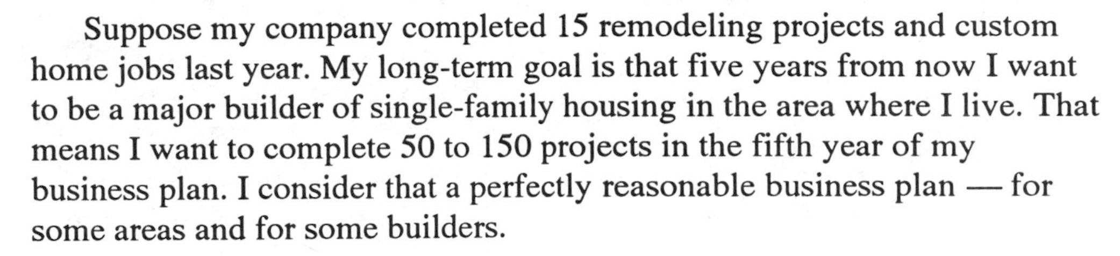

Suppose my company completed 15 remodeling projects and custom home jobs last year. My long-term goal is that five years from now I want to be a major builder of single-family housing in the area where I live. That means I want to complete 50 to 150 projects in the fifth year of my business plan. I consider that a perfectly reasonable business plan — for some areas and for some builders.

But let's also suppose I'm building in Connecticut. That probably makes my long-term goal unrealistic. Why? Because the "experts" predict zero population growth in New England and the Northeast over the next decade. If I'm sitting in the middle of a declining residential market, it's not realistic to believe I can multiply my business volume tenfold over the next five years. I can't expect results like that no matter how much time, effort and money I spend.

"You'll have to make changes. Adapt to existing conditions."

According to the experts, the United States will have a population of around 270 million by the year 2030. After that, population growth may stagnate, as it has in Europe for the last decade. Zero population growth isn't something we're accustomed to in this country. But it's possible anywhere.

When there's rapid population growth, contractors are busy with new construction. In a no-growth environment, repair, renovation and rebuilding will be more attractive, but custom home work may be unavailable. Another problem is that many of the most desirable areas of the country, the ones where people want to live, are already built out and are very expensive. What vacant land is available won't make suitable home sites for the people who want to buy.

Adapt Your Business to Your Market

You'll have to make some changes. Adapt to existing conditions. Find a new product that suits market conditions in your community. Learn how to build it profitably. What are your options? Some construction specialties will thrive in almost any economic climate.

Bridge repair work will draw lots of federal, state and local government money over the next decade, no matter what happens to our country or the economy. Bridges simply have to be fixed or they'll collapse. If you want to stay in Connecticut, consider specializing in bridge repair and maintenance. You'll be assured of a market.

Consider Renovation and Remodeling

You don't like bridge repair? OK. There are other options. Many contractors are getting into residential rehabilitation. They buy an older home for $150,000, do a major rehabilitation and offer it for sale at $250,000. Select carefully and you can make a bundle in a strong market. *Profits in Buying & Renovating Homes* by Lawrence Dworin is a good source of information on this specialty. Again, check the order form in the back of this book.

If you don't like the spec rehab business, consider custom remodeling. It's usually strongest in fully-developed communities where property values are rising and most homes are at least 20 years old.

Even if we reach zero population growth nationwide, some areas will continue to grow. The baby-boomers of the 1940s are approaching retirement age. Some plan to retire early. They'll move out of the big cities and cold climates to places where there's milder weather and a more relaxed lifestyle. If you're living and working in one of those areas, you'll continue to experience some growth, but maybe not at the rate we had in the 1980s. There will be more competition for the available jobs. Contractors will move along with the rest of the population. The construction industry will become even more volatile and competitive.

Go where the jobs are

In the 1970s it was very uncommon for small contractors to bid on jobs thousands or even hundreds of miles from their home community. Florida home builders didn't develop condos in Oklahoma. Only the largest engineering and construction companies built all over the country. Today, it's not unusual at all for even smaller contractors to look for work where the jobs are. That's why there are fewer and fewer little pockets of construction prosperity. Competition is more likely to be stiff all over.

Focus on a Specialty

There's been a long-term debate in the construction industry about whether it's better to specialize, or be a jack-of-all-trades. An industrial general contractor who bids only to the steel industry for plant expansion projects is a specialist. At the other extreme is the contractor who bids on commercial, industrial and residential projects from every developer and architect in town, regardless of the size or type of job.

"The specialist also has a better opportunity to establish a reputation for quality work . . ."

Market cycles tend to favor one type of contractor over another. In the tight market of the 1990s, the advantage is with the specialist. During the worst construction years, nine out of ten contractors are complaining of little work and no profit. But there's one in ten who's having a great year, and looking forward to a rosy future. Without exception, that tenth contractor is in a very specialized market. They have few competitors, know their business very well, and they keep at it.

The specialist also has a better opportunity to establish a reputation for quality work in that specialty. Specialists tend to work for a core of repeat customers who place their jobs on the basis of past performance, not necessarily the lowest price.

I conclude from this that contractors who focus their efforts and refine their specialized skills have a better chance of surviving when others don't. Leave diversification to others and to the good times when every builder has plenty of work.

The Rules of Supply and Demand Still Apply

Here's the point of this discussion about business planning: No company can ignore market conditions. Supply and demand affect both costs and markup. Your estimates have to consider the economic cycle. When markups are good, builders don't have to worry as much about accurate estimates. Cost estimates don't have to be accurate to 2 or 3 percent if there's a 30 or 40 percent markup in the job. You just figure your costs and add a reasonable markup. That's enough in a good market. But those conditions never last more than a few years.

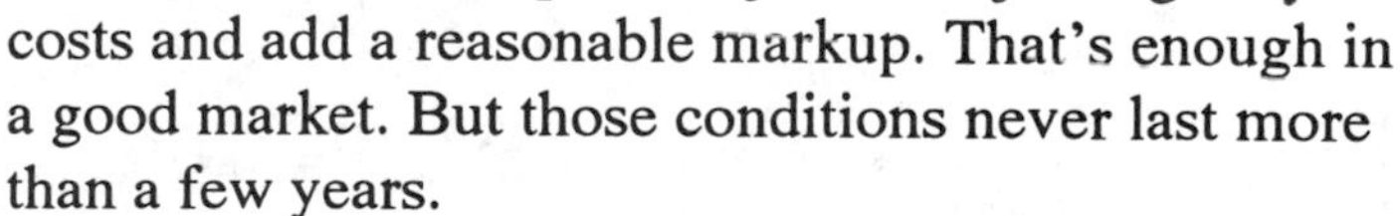

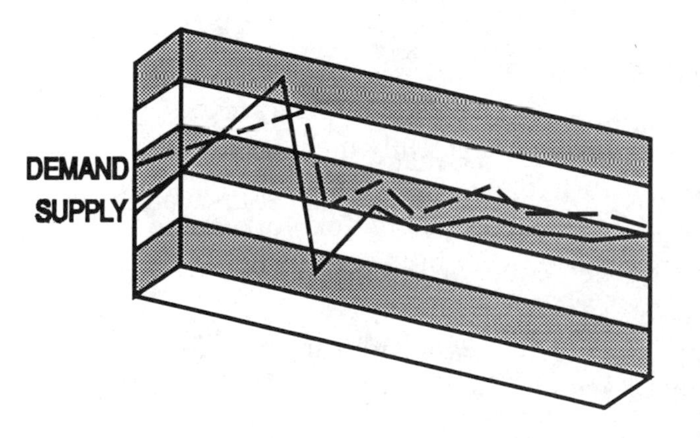

In the late 1960s we had EPA funding for sewer projects. The federal government was financing 95 to 100 percent of new sewers and sewer treatment projects through EPA bonds. That money ran out in 1970. But in 1968 and 1969 nearly any sewer or sewer treatment plant contractor could use a 50 percent markup and be reasonably sure of getting plenty of jobs. There was more work available than there were machines and people capable of handling it. Conditions like that don't last forever. Contractors who expect they will are in for a disappointment.

Markups fell pretty fast for sewer contractors in 1970. Sewer contracting became a high-risk business. Many firms went belly up. Others found ways to survive, probably because they had a business plan and

could cut overhead nearly as fast as business volume fell. Easy times for sewer contractors evaporated. But the bust lasted for only a few years. Then the market returned to normal — with a lot fewer competitors in business.

Who survived? Those who learned to be nimble. Those who could be more mobile and migrate to the communities where there was work. Those who could switch to the type of work that was needed at that time.

Profit As a Percentage of Sales and Net Worth

Competition in the construction industry tends to keep profit margins low compared to other industries. Dun and Bradstreet, the national credit reporting company, follows profit margins in manufacturing and construction industries. Dun and Bradstreet reports that profit margins for construction contractors and subcontractors average only 1.2 percent of sales in a typical year. That's less than one-half the average of all manufacturing businesses. Only 6 of 71 business categories have average profits less than the construction industry.

Even though profits measured as a percentage of sales tend to be low, the profit per dollar of net worth has traditionally been high. Dun and Bradstreet has reported that construction contractors average an after-tax profit of 9 percent of their net worth. This places construction seventh from the top of 71 manufacturing and construction industries. Profit per dollar invested is so high because fairly few dollars are needed to set up a construction contracting or subcontracting business.

Don't Just Sit Tight

Suppose business is scarce. You're saying, "We're doing OK now. We're bidding a lot of projects. We're not getting many, and we're not making any money on them. But we'll be here when things pick up again."

I think you're kidding yourself. Don't count on things getting better. They may not. Structure your business for existing (or even worse) conditions. Don't assume that the pressure will ease and margins will widen, or that there will suddenly be another flood of work.

Do extra planning for the slow times

Plan for the slow times. The good times will take care of themselves. Don't get over-extended. Avoid heavy commitments on fixed expenses like equipment and loans. Be aware that the business cycle rewards those who can anticipate it and punishes those who can't. Be one of the few who understands that no tree grows to the sky. Your market won't get stronger and stronger. No market does. Especially in the construction industry.

Are You in More Than One Business?

Most companies are. The imaginary company we'll use in our examples throughout this book is a general contracting firm which does its own carpentry and concrete work. That company is really three companies in one. They're a general contractor. They're also concrete and carpentry subcontractors. It doesn't matter that as subcontractors they work for only one customer, the general contractor. They're still in the carpentry and concrete subcontracting business.

If you're in the same situation, review the profit from each part of your business from time to time. Decide what's making money and what should get an increasing share of your attention and company assets. Eliminate or reduce any part of the business that doesn't fit into your long-term goal. Maybe most of your profit comes from one trade where your crews are both highly skilled and extremely competitive. Emphasize that and you'll become the most effective competitor in that specialty.

The Long-term Plan Is the Answer

That brings us back to the long-term business plan. Make a decision. Be certain that what you're estimating and building is your best bet for the long term. If not, make some changes.

But resist the temptation to make major changes in a five-year plan just because you expect a weak construction market next year. Instead, plan to tighten up your overhead when volume falls. Don't let a little ripple change the course of your ship. There will always be good years and bad years.

Make some guesses about conditions five or ten years down the road. Will there be substantially less demand for what you build? If so, reconsider your long-range plan. Make gradual changes that lead you into a more attractive market where there's likely to be a more comfortable markup.

Do a little planning now. It can make you a more effective estimator for many years to come.

Chapter 3
Getting Ready to Estimate

This chapter is an overview of what most experienced estimators know about construction cost estimating. I'm not going to get very specific in this chapter. That comes later. And there aren't any examples or practical exercises in this chapter. Again, that comes later. But don't get the idea that this chapter isn't important. It is. I'm going to assume that you've absorbed the basic information here before you go on to Chapter 4.

First, I'll talk about plan reading, what you can expect to see on the plans and specs, and what you should look for. Then I'll explain how estimators usually organize their estimates. There's a generally-accepted sequence that most estimators follow. I recommend that you follow it too.

Next, I'll touch briefly on estimating procedure for each of the major divisions of a construction estimate: site work, concrete, masonry, carpentry, finishes, and so on. The contractor's labor burden is an important part of your estimate. It's covered in this section, too.

Finally, I'll suggest ways to inoculate your estimates against costly mistakes. The fewer errors, the more likely you'll make a reasonable profit on the job.

It All Starts With Plans

In the remodeling business, you may see the job site before plans are drawn. That's the exception. Usually, your introduction to every estimating job will be a set of construction drawings and written specifications. Plans come first because the code requires it. You need a set of working drawings before you can get a permit down at your local building department office. The building department will stamp the plans and issue a permit for the work described in the plans.

The remodeling business is different. The cart goes before the horse. Sometimes the bid comes first. You prepare a bid to sell the job. You have to base your estimate on the measurements you take when you first see the job:

- Linear feet of cabinets to install
- Square feet of partition wall to remove and replace
- Square yards of flooring to lay

If your customer accepts your bid, then you or a designer will draw a set of plans to submit for approval at the building department.

Whether you begin with plans (for new construction) or a visit to the job site (for a remodeling project), the first step is to get a quick mental picture of the job. If you don't have plans, get that mental picture by standing in the room that's going to be remodeled. If you have plans, get the mental picture by reading the plans.

Avoid Preliminary Quotes

While you're standing in that room getting a mental picture and taking measurements, don't fall into a common trap. On a remodeling job, the owners are usually anxious about one question more than any other: "How much is it going to cost?" My advice is to avoid answering that question until you complete your estimate and bid. Quote a price that's too high, and you've lost the job before it's even bid. Give an evasive answer and you're considered dumb or dishonest. Quote a price lower than your actual bid and the owners will ask you to stick by the original (preliminary) quote.

"'How much is it going to cost?' My advice is to avoid answering that question until you complete your estimate and bid."

Instead, answer the "How much" question with another question. For example, say, "I've seen jobs like this done for as little as $18,000 to $20,000. How much do you want to spend?" When you've got an answer to that question, describe a similar job you know about. But be sure to stress differences between that job and the job you're going to recommend to these owners. Quote prices for other jobs, but don't ever offer a preliminary quote on the current job.

Courtesy: Merillat Industries, Inc.

Figure 3-1 Perspective drawings help sell a job

Emphasize that your bid will include exact prices and offer several options the owners might want to consider. Tell them you'd rather not quote a price until you complete the bid. Then, don't say anything more about the price.

Use Perspectives

In remodeling work, nothing sells a job better and faster than a perspective drawing of the finished kitchen, bath or room addition you propose. Prepared in pen and ink or in watercolor, the perspective drawing should show an elevation view of what the finished project will look like. While in no sense a working "plan," a perspective drawing will give you and your client an immediate impression of the finished product. In fact, I recommend you submit a perspective drawing with your bid any time you prepare an estimate for a homeowner.

Even in new construction, I recommend you begin your review of the plans with a perspective drawing, if there is one. Many larger projects include a perspective drawing to help the owner and lender visualize what the designer intends.

Review the Plans

If you have working drawings for the project, they'll include floor plans, elevation views, section views and whatever detail drawings are needed to show features not apparent on floor plans or elevation views. These sectional and detail drawings are important to the estimator. Generally, the more detail drawings, the more expensive a job will be. It's the details that run up construction costs.

Floor plan & elevation views

A floor plan shows what a building would look like from above if you lifted the roof off. It's a horizontal section cut through the walls, windows and doors to show their exact position and size. You'll have one floor plan for every floor in the building.

Elevation views show a building as though it were seen looking straight at it, instead of from above. You'll have at least one elevation view for each side of a building. Look at the elevation views to see the height of doors and windows and pitch of the roof. Information like that won't appear on floor plans.

Picture the planned building in your mind. Compare the plan for this building with others you've seen or built. Study the heights, lengths and widths. Compare the dimensions you see with dimensions you're familiar with in other buildings. Try to visualize the completed structure. Begin to think about problems you can foresee and challenges you'll have to overcome. This is the beginning of your construction cost estimate.

The Plan Package

A complete set of construction documents includes the following:

- Plans (floor plans, elevation views, detail views)
- Schedules (lists of fixtures, doors, windows, etc. to be installed)
- Job specifications
- The construction contract

Together, these should have nearly all the information you need to prepare a construction cost estimate.

Once you've scanned the plans and perspective drawings, if any, pick up the specifications. The specs have detailed information about materials to be used and instructions about procedures to follow during construction. They define the way the work is to be done and describe the quality of materials to be used — points not easily shown on a plan. Almost everything in the specifications will add to the cost of construction. Anything you miss or misread in the specs will probably reduce the profit you make on the job. That's why reading and understanding the specs is so important.

"C" sheets

The plans, elevations and detail drawings probably come to you as a package bound on one edge. The order of binding depends on the architect's preference. Larger jobs will probably have a plot plan that shows the position of the building on the site. It also shows the location of sewer, gas, electric and water lines, distances to lot boundaries and details about excavation, drainage and parking spaces. Plans showing details like these are often called the civil drawings or "C" sheets.

The Architectural Sheets

"A" sheets

Below the "C" sheets will be the "A" sheets. These are the building plans drawn by an architect. Some architects place the elevation views first. The floor plans and details follow. Other architects feel the floor plans are more important and put them on top, beginning with the basement plan (if any), the first floor plan, the second floor plan, and so on. The elevation views follow the floor plans.

Next, detail views and schedules of purchased and installed architectural items will probably follow. For example, you'll find a door schedule and a window schedule here. These list the size, type and style of each window and door shown on the plans.

Then you'll find the plans prepared for use by subcontractors such as the electrical, plumbing, HVAC and landscaping subs. On some jobs, subcontractors will prepare their own plans. If that's the case, you won't find any mechanical or electrical drawings with the architectural plans. Instead, the architect prepares an outline spec that indicates only the quality of fixtures and equipment required. The subcontractor prepares working drawings when he gets the contract.

Follow a logical sequence

The order you follow in reading the plans depends on your experience and preference. My best advice is that you be systematic and consistent. Follow any logical sequence, *but stick to it.* Don't get into the habit of jumping from one sheet to another. If you read a little here and a little there, sooner or later you'll miss an important item from your bid. When you have to pay for that item out of your profits, or your pocket, you'll better appreciate the importance of following a logical sequence.

The Basement Plan

If there's a basement, I recommend you start with the basement plan. I like to work from bottom to top. That's the way you build the building, so that seems most logical to me. True, I have to refer to other plans and elevations occasionally to understand what's required for basement construction. But I'm going to find out everything I can about the basement before I move on.

Nearly all basements have certain things in common. You'll see two parallel foundation lines that form a somewhat irregular rectangle. The outer line represents the outside of the foundation wall. The inner line represents the inside line of the foundation. Note the distance between opposite walls. Are wall dimensions measured from the inside of one wall to the inside of the opposite wall? Or are they measured from the outside of one wall to the outside of the opposite wall? Maybe they're measured from wall center to wall center. How thick is the foundation wall? While you check these dimensions, add up the interior measurements to see if the total checks with overall lengths. If there's an error somewhere in the plans, find it early before you do too many calculations.

Don't crow about mistakes

While you compare plan sheets, check figures, and scale off dimensions, you'll often find an error. Something doesn't add up or is obviously wrong. Sometimes you can fix the mistake yourself, without calling the architect. In fact, I recommend you don't call the architect just to boast that you've discovered an error in the plans. Crowing about a minor mistake is likely to do more harm than good. It irritates and embarrasses the architect, and it tends to get back to your competition. Let them find their own mistakes. If you see an error, and you're certain it's an error, and can correct it, make that correction yourself. Of course, if you're not absolutely certain what the architect intends, there's no choice but to ask for more information.

Study the foundation walls further, and notice the footings. If they're not shown on the basement floor plan, you'll find this information in a detail drawing on some other sheet.

Check the elevation views to find the height of the finish grade on the foundation wall. Look for basement details such as a laundry room, drainage outlets, layouts for heating and air conditioning equipment, fuel storage, closets, tubs, toilets, or a platform for equipment. Examine the placement of supporting columns and piers for stairs. What's the height of the foundation wall? What type and how many basement windows are shown? Are there any partitions on the elevation views?

When you've got a good impression of the basement, go on to the first floor plan.

The First Floor Plan

As with the basement, begin with a general impression. Then get down to the specifics. Look at dimensions, cabinets, closets, fixtures, and decorative finish carpentry. Items like these will probably appear in detail drawings on a larger scale and on a separate sheet.

Windows and doors appear on floor plans as two parallel lines. Check the location of each. A reference letter on the floor plan references each window and door to a separate schedule. Are the windows grouped and separated by a mullion? You may have to refer to the elevation views to learn this.

Next, examine the stairs. Check the riser height and tread width. Note any landings. Are these simple box stairs framed between two walls? Or is it an open stairway requiring posts, rails and balusters? Take careful note, too, of stairs leading down to the basement. Is it a single flight with a straight run or do you have to frame a turn and a landing? The cost differential is significant.

Pay special attention to the fireplace, chimney and flues. Flues should show clearly on the plan. If a fireplace is indicated, you'll find a detail drawing that shows the face brick, lining, hearth, and ash pit required — unless the masonry subcontractor is expected to prepare these drawings. You may find notes on the plan near the fireplace that refer to detail drawings of a seat, bookcase or fireplace form that's to be built into the masonry.

Finally, take a quick look at the electric and mechanical sheets, if there are any. Here you'll see the electric outlets, service panel, plumbing and heating equipment. That usually completes your evaluation of the interior first floor plan.

The Second Floor Plan

Now go on to the second floor plan, if there is a second floor. Here you'll find the same outline as on the first floor. Any exceptions will probably be indicated with a dotted line to show the variation from the first floor perimeter.

Again, check the windows and doors. In a residential building, you'll probably find at least one bathroom with bathtub on the second floor. The conventional way of showing these rarely varies from one architect to another.

Note the continuation of the chimney and flues through the second floor. Your plan may also show the front porch roof and the type of roof cover.

The Elevations

When you've examined all the floor plans, go on to the elevation views. There will be at least four: front, rear and two sides. Each represents what you would see if you were looking directly at that face. You'll see all the door and window heights and widths, the shape of the roof, its slope, the style of cornice, porches, balustrades and outside trim. Notes on elevation views should describe materials used for wall and roof coverings. Not all dimensions will be shown. For example, you may have to scale off on the elevation views the exact location of windows, vents and louvers.

The first floor exterior elevation views will show steps, porches, posts, a covered entryway and other exterior details included in the plan. There may be a large-scale drawing of the front entry or a living room bay.

Large-scale detail drawings may be included to show the style and construction of cornices, porches and bay windows. Section lines drawn through a wall or roof indicate that a detail drawing is included in the plans. This detail view will have the information you need to build (and estimate) what the architect intends.

This preliminary look at the plans should give you the big picture — now you know what you have to estimate. You haven't done any estimating yet, but at least you know what's required. Maybe you've also come across a few mistakes or items that have to be cleared up before you can begin.

Examine the Site

On any project, I recommend that you visit the site sometime before you complete the estimate. When you make that visit, look for anything that may have an impact on costs.

Site factors that impact costs

- What's the soil like?
- Where will we store materials?
- Will access be a problem in wet or cold weather?
- How far are runs for temporary electric and water supply?
- Can concrete trucks chute directly into the forms?
- How close are the neighbors?
- What site clearing will be required?
- Will power lines obstruct the work?

Learn what you can about the subsoil. Is there rock below the surface? How high is the water table? In all excavation work, there's always an opportunity for surprises. Will the owner supply core samples? Or will you or your subcontractor have to pay for samples?

Generally, the law allows you to recover the additional cost if unexpected conditions increase the cost of excavation. But many unexpected costs are just your bad luck. There may be no way to recover for some extra expenses. For example, suppose surface water from a nearby drainage channel floods the project during the rainy season. Who's going to pay to have the site dried out? The contractor, of course. That's the type of problem your site visit should alert you to.

The Cost Data File

I believe the best estimators are systematic, consistent and orderly. If you aren't well organized by nature, at least be organized in your approach to estimating. Have a system that you follow on every estimate. I recommend you estimate the work in the same order you'll build the job.

For example, the first task is usually preparing the site. Then comes excavation, then foundations, exterior walls, etc. — finishing up with decorating, hardware and painting. Omit something important and you'll probably notice it immediately if you estimate in the same order as you do the work.

Develop a practical filing system

I also recommend that you keep a cost estimating notebook or have some other way of collecting and organizing estimating data. Probably the most convenient and practical filing system for this purpose is a simple loose-leaf notebook with 8½ x 11-inch pages. You'll need a number of sheets for each estimate. But all sheets of the same estimate should be filed together (you can use tabbed dividers between estimates) and have the same job number in the upper right-hand corner. The cover sheet should bear the job address, the owner's and architect's names, addresses and phone numbers, the bid date, the name of the estimator and the name of the estimate checker.

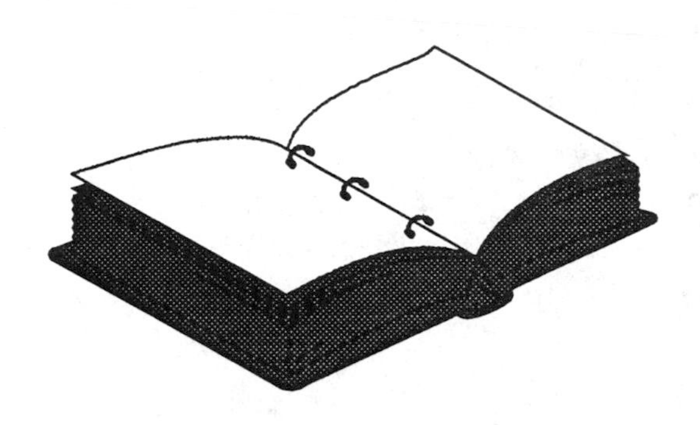

With a loose-leaf notebook, it's easy to lift all the sheets for any job out of the book. When your book is full or when an estimate is rejected, remove those pages and file them together in job number order in a file drawer. Keep an index of job names and job numbers in your notebook or in a card file. That makes it easy to locate older estimates when you need them.

You'll refer to the cost data in your estimating notebook many times while you compile new estimates. As you accumulate more and more completed estimates, this file will become your most valuable and most reliable estimating resource.

Recording Job Costs

Of course, the most valuable information in your estimating notebook will be the cost records of already-completed jobs. These are the best possible guide to costs on future jobs. Understand that there are two possible profits in every job. One ends up in the company bank account. The other is what you learn about estimating costs on future jobs. Don't waste that second profit. Record actual costs and actual crew productivity on every job. In Chapter 8 I'll offer an easy, simple, quick, almost painless way to record job costs and crew productivity. All that information should go into your estimating notebook.

"Understand that there are two possible profits in every job."

On every job you have, record actual costs as carefully as you recorded estimated costs. Either leave a column on your estimates for actual costs or keep actual costs on a separate sheet that you can file in your notebook behind the cost estimate pages.

Using Cost Guides

Of course, many published cost guides are available. If you don't have actual costs, you may have to rely on published costs. But at best, these cost guides are only second-hand information. No cost guide is a substitute for costs quoted by your local building material dealers. None are based on *your* crews, working under *your* supervisors, and on *your* type of work.

Don't rely on the opinion of others when you compile estimates. Keep your own costs and organize them neatly in your estimating notebook.

Using Your Own Cost Data

The time it takes to shingle a roof, lay flooring tile, set an interior door, lay a thousand brick — these are all notes you should record for ready reference in your estimating notebook. Records like these help you both anticipate costs and identify productivity problems before they turn a good job into a major loss.

For example, suppose your cost data shows that two carpenters can frame a certain type of stairway in 10 manhours. Suppose also that you estimate 10 hours, and it actually takes two workers two full days — 16 manhours. Something is wrong. Either the job was very different or the tradesmen weren't working efficiently. If it was very different, did the job include extras you could bill to the owner? If productivity was poor, what can you do to be sure it doesn't happen again?

Keep in mind that productivity (manhours) varies widely from crew to crew, from job to job and from day to day. For example, brick masons working on a sunny spring day get more done than they do when a cold wind makes hands stiff and bulky clothing restricts movement. That doesn't mean there's no way to estimate productivity. It means you have to record working conditions when those conditions affect productivity. At least, add a note to any productivity figures you record in rainy or cold weather.

Doing the Take-off

As I suggested before, the most convenient way to organize your estimates is in an 8½ x 11-inch loose-leaf notebook. Lined pages like you used in high school are fine. Or you may want to buy estimating forms sold by some publishers. In any case, identify each estimate with a number. Put that number on every sheet of the estimate. And number the pages in each estimate consecutively. There's more on this subject in Chapter 5.

This bears repeating: When you begin your quantity take-off, list materials in the same order you'll do the work. That reduces the chance of a major omission, the estimator's worst nightmare. Your estimate for particular items on the job may be a little too high or a little too low. But with any luck, your high and low estimates will balance out. But the estimate on anything you forget is always zero. That's a 100 percent miss. Make very many of those, and even the best job can become a solid loser.

Group costs by section, such as concrete, finish carpentry or roofing. Include a subtotal for each section of your estimate. Prepare your estimates in distinct sections. That makes it easier to correct errors or change prices when necessary. Changing a cost in any section affects only that section. Just accumulate the subtotals again to find the new grand total.

Classification of Operations

The Construction Specifications Institute has developed a classification system called the Masterformat. You can get a copy of the current Masterformat for $60 from:

Construction Specifications Institute
601 Madison Street
Alexandria, VA 22314
Phone 703-684-0300, FAX 703-684-0456

In Canada, write to:

Construction Specifications Canada
1 St. Clair Avenue West, Suite 1206
Toronto, Ontario, M4V 1K6

Written specifications for most large jobs follow the Masterformat.

The Masterformat places all construction activities into 16 major headings. Each major heading has many subheadings. See Figure 3-2. Even if you don't follow the Masterformat system precisely, I recommend that you get familiar with this system of organization. Specifications will be in Masterformat order. It's easier to find what you're looking for in the specs if you know which section covers each subject.

Compiling an Estimate

In Chapters 5 and 6 you'll get some hands-on practice doing a material and labor take-off from a set of plans. In this chapter I'm going to sketch the estimating process in general. Later we'll get down to specifics.

Regardless of the type of work, the quantity take-off procedure is nearly always the same:

The quantity take-off procedure

- Analyze the work to be done.
- Consider the time required, the materials needed and the equipment necessary to perform each operation.
- Picture in your mind how each task has to be done.

Division 1 — General Requirements
01010 Summary of work
01020 Allowances
01025 Measurement and payment
01030 Alternates / alternatives
01035 Modification procedures
01040 Coordination
01050 Field engineering
01060 Regulatory requirements
01070 Identification systems
01090 References
01100 Special project procedures
01200 Project meetings
01300 Submittals
01400 Quality control
01500 Construction facilities
01600 Material and equipment
01650 Facility startup / commissioning
01700 Contract closeout
01800 Maintenance

Division 2 —Sitework
02010 Subsurface investigation
02050 Demolition
02100 Site preparation
02140 Dewatering
02150 Shoring and underpinning
02160 Excavation support systems
02170 Cofferdams
02200 Earthwork
02300 Tunneling
02350 Piles and caissons
02450 Railroad work
02480 Marine work
02500 Paving and surfacing
02600 Utility piping materials
02660 Water distribution
02680 Fuel and steam distribution
02700 Sewerage and drainage
02760 Restoration of underground pipe
02770 Ponds and reservoirs
02780 Power and communications
02800 Site improvements
02900 Landscaping

Division 3 — Concrete
03100 Concrete formwork
03200 Concrete reinforcement
03250 Concrete accessories
03300 Cast-in-place concrete
03370 Concrete curing
03400 Precast concrete
03500 Cementitious decks and toppings
03600 Grout
03700 Concrete restoration and cleaning
03800 Mass concrete

Division 4 — Masonry
04100 Mortar and masonry grout
04150 Masonry accessories
04200 Unit masonry
04400 Stone
04500 Masonry restoration and cleaning
04550 Refractories
04600 Corrosion resistant masonry
04700 Simulated masonry

Division 5 — Metals
05010 Metal materials
05030 Metal coatings
05050 Metal fastening
05100 Structural metal framing
05200 Metal joists
05300 Metal decking
05400 Cold formed metal framing
05500 Metal fabrications
05580 Sheet metal fabrications
05700 Ornamental metal
05800 Expansion control
05900 Hydraulic structures

Division 6 — Wood and plastics
06050 Fasteners and adhesives
06100 Rough carpentry
06130 Heavy timber construction
06150 Wood and metal systems
06170 Prefabricated structural wood
06200 Finish carpentry
06300 Wood treatment
06400 Architectural woodwork
06500 Structural plastics
06600 Plastic fabrications
06650 Solid polymer fabrications

Division 7 — Thermal and moisture protection
07100 Waterproofing
07150 Dampproofing
07180 Water repellents
07190 Vapor retarders
07195 Air barriers
07200 Insulation
07240 Exterior insulation and finish systems
07250 Fireproofing
07270 Firestopping
07300 Shingles and roofing tiles
07400 Manufactured roofing and siding
07480 Exterior wall assemblies
07500 Membrane roofing
07570 Traffic coatings
07600 Flashing and sheet metal
07700 Roof specialties and accessories
07800 Skylights
07900 Joint sealers

Division 8 — Doors and windows
08100 Metal doors and frames
08200 Wood and plastic doors
08250 Door opening assemblies
08300 Special doors
08400 Entrances and storefronts
08500 Metal windows
08600 Wood and plastic windows
08650 Special windows
08700 Hardware
08800 Glazing
08900 Glazed curtain walls

Division 9 — Finishes
09100 Metal support systems
09200 Lath and plaster
09250 Gypsum board
09300 Tile
09400 Terrazzo
09450 Stone facing
09500 Acoustical treatment
09540 Special wall surfaces
09545 Special ceiling surfaces
09550 Wood flooring
09600 Stone flooring
09630 Unit masonry flooring
09650 Resilient flooring
09680 Carpet
09700 Special flooring
09780 Floor treatment
09800 Special coatings
09900 Painting
09950 Wall coverings

Division 10 — Specialties
10100 Visual display boards
10150 Compartments and cubicles
10200 Louvers and vents
10240 Grilles and screens
10250 Service wall systems
10260 Wall and corner guards
10270 Access flooring
10290 Pest control
10300 Fireplaces and stoves
10340 Manufactured exterior specialties
10350 Flagpoles
10400 Identifying devices
10450 Pedestrian control devices
10500 Lockers
10520 Fire protection specialties
10530 Protective covers
10550 Postal specialties
10600 Partitions
10650 Operable partitions
10670 Storage shelving
10700 Exterior protection devices
10750 Telephone specialties
10800 Toilet and bath accessories
10880 Scales
10900 Wardrobe and closet specialties

Division 11 — Equipment
11010 Maintenance equipment
11020 Security and vault equipment
11030 Teller and service equipment
11040 Ecclesiastical equipment
11050 Library equipment
11060 Theater and stage equipment
11070 Instrumental equipment
11080 Registration equipment
11090 Checkroom equipment

Figure 3-2 The Masterformat system

11100 Mercantile equipment
11110 Commercial laundry and dry cleaning
11120 Vending equipment
11130 Audio-visual equipment
11140 Vehicle service equipment
11150 Parking control equipment
11160 Loading dock equipment
11170 Solid waste handling equipment
11190 Detention equipment
11200 Water supply and treatment equipment
11280 Hydraulic gates and valves
11300 Fluid waste treatment and disposal
11400 Food service equipment
11450 Residential equipment
11460 Unit kitchens
11470 Darkroom equipment
11480 Athletic, recreational and therapeutic
11500 Industrial and process equipment
11600 Laboratory equipment
11650 Planetarium equipment
11660 Observatory equipment
11680 Office equipment
11700 Medical equipment
11780 Mortuary equipment
11850 Navigation equipment
11870 Agricultural equipment

Division 12 — Furnishings
12050 Fabrics
12100 Artwork
12300 Manufactured casework
12500 Window treatment
12600 Furniture and accessories
12670 Rugs and mats
12700 Multiple seating
12800 Interior plants and planters

Division 13 — Special construction
13010 Air supported structures
13020 Integrated assemblies
13030 Special purpose rooms
13080 Sound, vibration and seismic control
13090 Radiation protection
13100 Nuclear reactors
13120 Pre-engineered structures
13150 Aquatic facilities
13175 Ice rinks
13180 Site constructed incinerators
13185 Kennels and animal shelters
13200 Liquid and gas storage tanks
13220 Filter underdrains and media
13230 Digester covers and appurtenances
13240 Oxygenation systems
13260 Sludge conditioning systems
13300 Utility control systems
13400 Industrial and process control systems
13500 Recording instrumentation
13550 Transportation control instrumentation
13600 Solar energy systems
13700 Wind energy systems
13750 Cogeneration systems
13800 Building automation systems
13900 Fire suppression and supervisory systems
13950 Special security construction

Division 14 — Conveying systems
14100 Dumbwaiters
14200 Elevators
14300 Escalators and moving walks
14400 Lifts
14500 Material handling systems
14600 Hoists and cranes
14700 Turntables
14800 Scaffolding
14900 Transportation systems

Division 15 — Mechanical
15050 Basic mechanical materials and methods
15250 Mechanical insulation
15300 Fire protection
15400 Plumbing
15500 Heating, ventilating, and air conditioning
15550 Heat generation
15650 Refrigeration
15750 Heat transfer
15850 Air handling
15880 Air distribution
15950 Controls
15990 Testing, adjusting, and balancing

Division 16 — Electrical
16050 Basic electrical materials and methods
16200 Power generation built-up systems
16300 Medium voltage distribution
16400 Service and distribution
16500 Lighting
16600 Special systems
16700 Communications
16850 Electric resistance heating
16900 Controls
16950 Testing

Figure 3-2 (cont.) The Masterformat system

It's essential that you're familiar with the work involved. You don't have to know how to do each operation yourself, but you should be able to make an educated guess at how many manhours thc job will take, how much material you'll need, and what equipment will do the job best.

Let's suppose the project is a two-story home with a basement. You've spread the floor plans out on the table in front of you. You put the foundation plan on top, the first floor plan next, the second floor plan below that, and so on. The specifications are also on the table.

From your site visit, you know that the site is a quarter-acre covered with light brush and two trees with trunks about 16 inches in diameter. The trees and brush have to be cleared before you can begin construction. Your estimate is that two men using hand tools will require about 6 hours to cut down and trim the trees and remove the stumps. You allow another 2 hours for the crew of two to clear the brush. So two men working with hand tools and a truck should be able to clear the site in one day (16 manhours). Record that estimate on your take-off sheet under "Site Clearing."

Next comes the survey and the erection of batterboards and lines. A crew of two carpenters and one laborer will lay out the foundation from markers set by the surveyor. The surveyor's fee will be $450. The carpenters will require about half a day for their layout, and will use about 100 board feet of lumber and stakes. You record the survey cost as a subcontract item and the layout time as 12 hours, using 100 board feet of lumber. Next comes the site work.

Site Work

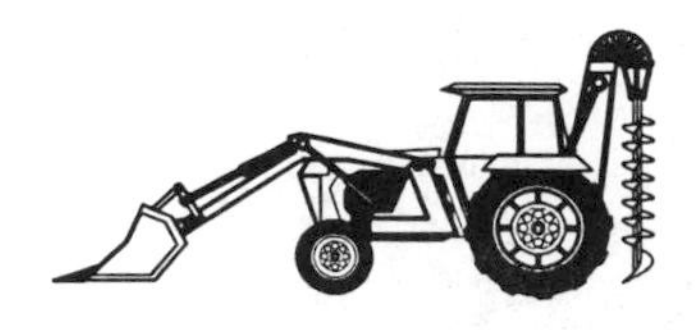

You determine excavation costs by the quantity of soil moved, the type of soil, the equipment required, and the distance the soil has to be transported. If soil has to be imported or exported, the cost will run much higher than on a balanced job where no soil has to be moved on or off the site. For this job, specifications permit us to excavate the basement and spread the soil on the site.

If the water table is high or if excavation is done in the rainy season, you may need a pump to dry out the basement. Include the labor required for pumping, rental on the pump, and cost of fuel.

Cost of the basement drain comes next. Consider the depth of the pipe trench, type of soil, linear feet of pipe of each size, and the pipe cost. For a French drain, figure the depth below the pipe inlet, the diameter, and the cost of stone fill. Figure excavation by the cubic yard, the delivered cost of crushed stone fill, and the cost to spread the excavated soil on site. Allow about one hour to lay out the drainage pipe and the French drain.

Concrete

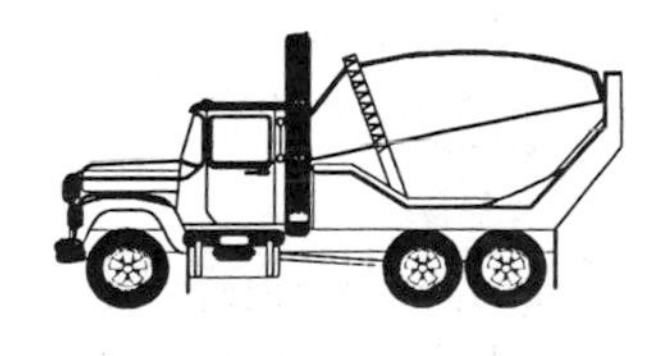

Start at one corner of the building to take off the footings, then work around the outside. Measure all cross walls in each direction, isolated piers, chimney footings, etc. Multiply the depth of the footing by the width by the length (all in feet and decimals of a foot). The result is the cubic feet of concrete for footings and piers. Then divide by 27 to find the cubic yards required. Add about 10 percent for waste and over-excavation.

For concrete flatwork, figure the square feet at each thickness (in decimals of a foot). Divide by 27 to convert cubic feet to cubic yards of concrete. Add about 5 percent to the calculated volume for waste. Add the labor cost to form, pour and finish the concrete.

Masonry

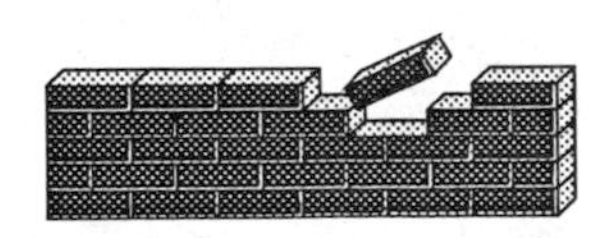

Your masonry materials dealer can supply reference tables that show how many brick or block you need per square foot of wall. Figure the square feet of wall and then multiply by the units per square foot to find the number of brick or block needed. To find the cost of brick or block, multiply the delivered cost per thousand by the number required in thousands. Add the cost of staging for masons, the cost of mortar, the cost of carrying brick to the job and the cost of laying brick.

Metals

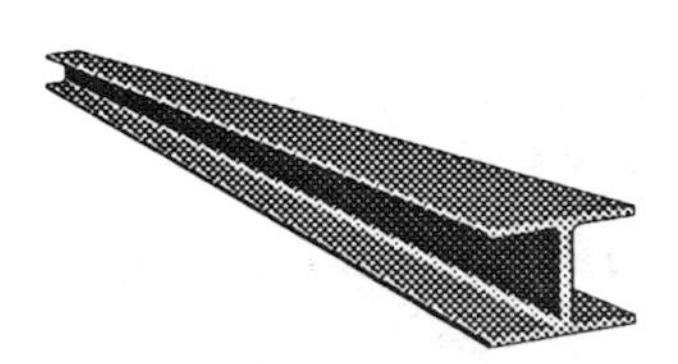

Very few residential buildings require structural steel. If a steel beam is required, a steel contractor will probably quote a price per pound for both erecting and painting the steel. Most structural steel has to be set with a crane. Iron work such as bolts, truss rods, straps and anchors are usually figured by counting the number required.

Wood and Plastic

Most framing is estimated by the board foot of lumber. A board foot is the quantity of lumber in a board that's nominally (by name) 1 inch thick, 1 foot long and 1 foot wide. A 2 x 6 has 1 board foot of lumber per linear foot. A 2 x 4 has 0.667 board feet per linear foot. To find the board feet per linear foot, multiply the thickness in inches by the width in inches. Then divide by 12.

For example, find how many board feet of lumber there are in a 2 x 8 that's 10 feet long. Here's how to calculate the answer.

Calculating board feet

- 2 (the thickness) times 8 (the width) is 16.
  16 divided by 12 is 1.333 board feet per linear foot.
- If there are 1.333 board feet in 1 linear foot, 10 linear feet have 10 times that, or 13.33 board feet.

List the linear feet for each size of lumber on your take-off sheet. Then estimate the labor required for framing.

Count the number of windows and doors for each size and style. List these on your take-off sheet, using information in the window and door schedules to identify each style.

You estimate molding and trim by the linear foot of room perimeter for each size and style. Be sure to deduct for wall openings such as doorways.

Spiral stairs, folding attic stairs and circular stairs are usually fabricated off the site. Experienced carpenters can build ordinary wood stairs on site. Be sure to include stair rails, newels, and balusters in your estimate when the plans require them.

Thermal and Moisture Protection

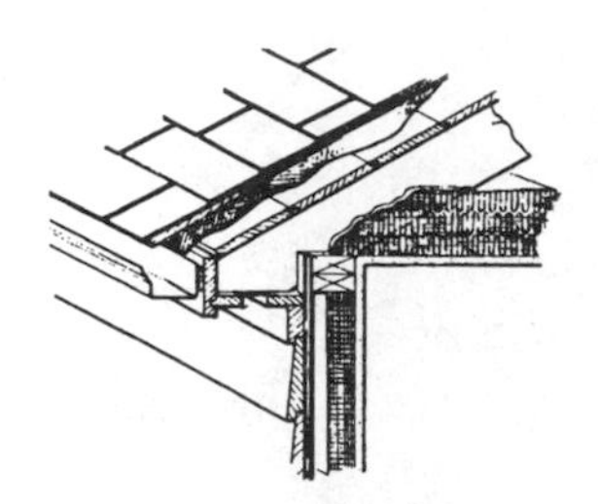

You figure insulation by square foot of wall, floor or attic covered. Batt and blanket insulation fills the wall cavity between studs and is made to fit between studs placed either 16 or 24 inches on center.

Roofing is estimated by the *square*, which is 100 square feet of roof surface. Wood or composition shingles and built-up roofing are the most common on residential buildings. As a general rule, figure the area to be covered, but ignore small roof openings such as skylights.

Finishes

Many general contractors use subs for most finish work, such as painting, drywall, flooring and tile. Calculate the area to be covered by each type of material. Small jobs, small rooms and small surfaces will usually require more time per square foot covered, and more waste allowance for materials.

Plumbing and Electrical Work

Estimate plumbing by the fixture, including piping, waste and vent stock. You can also figure electric wiring per outlet, fixture or switch. Because the length of piping and electric runs are about the same in most residential buildings, plumbing and electrical estimators often don't calculate the length of pipe, wire or conduit required. It's enough to know the number and quality of fixtures. In commercial and industrial buildings, where runs could be significant, record the length of every run of wire, conduit, cable and pipe.

The Contractor's Labor Burden

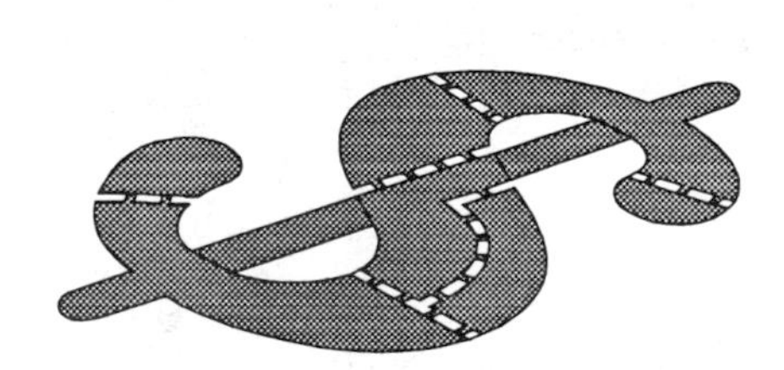

When you've recorded material quantities and manhour estimates on your take-off form, it's time to enter labor costs. For every contractor and on every job, your labor cost is not the same as the wages you pay your workers. You have a *labor burden* that adds between 25 and 35 percent to the labor cost. For every dollar of payroll, you have to pay an additional 25 cents to 35 cents in taxes and insurance to government agencies and insurance carriers. Many contractors routinely add 30 percent to the estimated labor cost to cover taxes and insurance. Here's where that money goes:

Unemployment insurance — All states levy an unemployment insurance tax on employers based on total payroll for each calendar quarter. The actual tax percentage is usually based on the employer's (and the state's) history of unemployment claims and may vary from less than 2 percent of payroll to 4 percent or more.

The Federal government also levies an unemployment insurance tax (F.U.T.A.) based on payroll. The tax at the time of this writing is about 0.8 percent of payroll.

Federal employment taxes — The Federal government also collects Social Security (F.I.C.A.) and Medicare taxes. Together, these come to about 7 percent of payroll, depending on employee earnings. Employers deposit these payroll taxes at least quarterly, depending on the total wages they pay.

Workers' Comp and liability insurance — All states require employers to maintain Workers' Compensation insurance coverage for their employees. If an employee is hurt on the job, coverage provides a supplementary income, medical treatment and rehabilitation therapy, if needed. Heavy penalties are imposed on employers who fail to provide the required coverage. The cost of Workers' Compensation insurance is a percentage of payroll that varies with the type of work each employee does.

Clerical and office workers have a very low Workers' Comp rate, usually about 1 percent of payroll. Hazardous occupations, such as pneumatic tunneling, may carry a rate of 50 percent of payroll or more. Most light construction trades have a rate between 10 and 15 percent, though the rate for roofers is usually about 35 percent of payroll for that trade.

The cost of Workers' Compensation insurance varies from state to state and from year to year, and depends on an employer's loss experience. Your insurance carrier can quote current figures for the type of work your employees handle.

Liability insurance — Every contractor should maintain liability insurance to protect the business from suit in the event of an accident. Liability insurance is also based on the total payroll and is usually about 3 percent of payroll. Higher liability limits cost more.

Here's a summary of the contractor's labor burden. The percentages listed are approximate, of course. Your accountant and insurance carrier can furnish accurate figures.

State unemployment insurance	4.0%
F.I.C.A. and Medicare	7.65%
F.U.T.A.	0.8%
Workers' Compensation	15.0%
Liability insurance	4.0%
Total contractor burden	31.45%

Every contractor who has a payroll is well-advised to add this burden into the labor cost on every estimate. And be sure to make the insurance and tax deposits when they're due. No contractor can ignore these requirements and operate for long.

Overhead

Many job costs are neither installation labor nor installed materials. For example:

Types of overhead

- Fire insurance
- Surety bonds
- Building permit
- Sidewalk permit
- Telephone
- Water
- Electricity
- Sewer and water connection fee
- Timekeeper
- Watchman
- Temporary office
- Repairs to adjoining property
- Job toilets

Project vs. company overhead

These costs are usually called *project overhead* because they're not labor or material costs.

Don't confuse *project* overhead with *company* overhead. General business expenses (like office rent, utilities and accounting fees) are different. Expenses like these go on about the same no matter how many or how few jobs you have. We'll devote most of Chapter 9 to a discussion of company overhead and talk more about project overhead in the next chapter.

Contingency and Escalation

Most estimators add a small amount to their bid to allow for *contingencies*, those unanticipated conditions that may develop during the course of construction. Work is seldom done faster than planned and problems often occur that make the work cost more. The right amount to add for contingency depends on the contractor and the job. However, for most light construction, around 2 percent is about right. Remodeling, repair work and excavation require a larger allowance because there are more unknowns when you begin work. You just can't accurately forecast all the problems before you start.

Escalation is the increase in labor, material and equipment costs between the time you prepare the estimate and the time you have to pay the bills. For example, the price of framing lumber when you estimate the job may not be the same as the price you pay when you order the lumber. Price changes for lumber, plywood and some metals can occur weekly, and prices may fluctuate wildly. The price of materials such as wallboard, concrete and glass are relatively stable. Changes are less frequent and by a lower percentage. If you can't get firm quotes for materials to be delivered in the future, either allow for price escalation, or exclude price increases from your bid.

The magazine *Engineering News Record*, published by the McGraw-Hill Book Company, 1221 Avenue of the Americas, New York, New York 10020, follows price trends in construction materials labor and equipment and can help you track price movements. *Construction Review*, published by the United States Government Printing Office, Washington D.C. 20402, follows prices of major groups of construction materials. Either of these publications will help you stay posted on price changes.

Profit

You're in business to make money — that's the profit in every job. Profit may be the last item on your estimate, but it's certainly not the least important. In Chapters 10, 11 and 12 we'll focus on profit: How to adjust your profit percentage to maximize company revenue.

Checking Estimates

Accuracy is essential in this business. Every experienced estimator knows that. And they also know that everyone makes mistakes. So what do you do to keep errors out of your estimates?

I've found only one effective way to get the errors out: Careful checking. It's not cheap. It's not quick. It's tedious. *But it's essential.*

"I've found only one effective way to get the errors out: Careful checking."

I'm going to talk more about estimate checking in the next chapter. But this is an important topic that deserves more than lip service. So I'll introduce the topic in this chapter and get into more detail in the next.

Underestimates and overestimates are common in this business. Your estimated cost may be 10 percent too low on one thing and 15 or 20 percent too high on something else. Accidents like that happen. Every estimator will admit it. But with a little luck, the underestimates will balance with the overestimates, resulting in a bid that's right on the money. Too often, that's an accident too, maybe more often that we're willing to admit.

But make your estimating system fail-safe when it comes to omitting things. Nothing will balance that out.

Everybody makes mistakes now and then. I've made my share — and I expect you will too, no matter how carefully you work. Once I checked an entire estimate carefully four times, looking for a mistake. "No," I thought, "it's perfect." Still, when I had an assistant check it, the mistake was so obvious that I couldn't believe it. In effect, I had become blind to the mistake no matter how many times I reviewed the estimate.

Nobody's perfect, construction cost estimators included. You're going to make mistakes. But there are ways to both reduce the opportunity for errors and catch the mistakes you make — before they lead you to a financial disaster.

The surest way to verify an estimate is to have two estimators work up costs for the same job. Then compare the totals. Where they disagree by more than a few percent, there's probably a mistake.

But double-figuring every job is a luxury most estimating offices can't afford because it doubles estimating cost. There's another way that costs less and works almost as well.

First of all, be consistent when you estimate. Develop a sequence and procedure that you like and that produces good results. Then follow it every time. In the next chapter, I'll explain the procedure I use.

Second, be sure your estimates are easy to check. Show all the steps you followed and the calculations you made. Leave your scratch sheets and calculator tapes attached to the estimate package. It should be clear from the estimate exactly how you developed each quantity and cost.

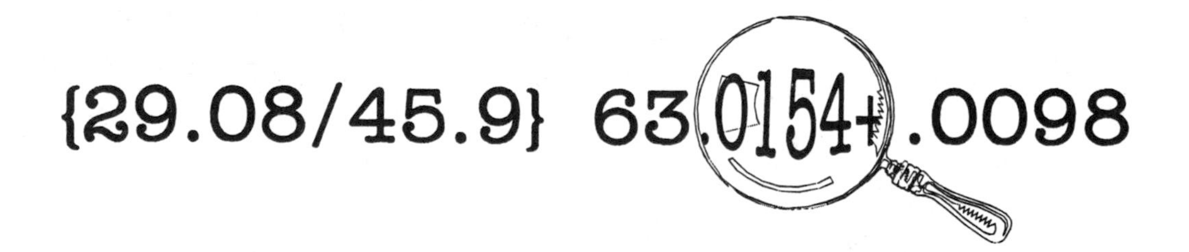

Third, never trust your own totals until you've checked them. Then have a second estimator review the figures. This is good work for an apprentice estimator. As an absolute minimum, have a clerk check the math and verify that you've transcribed figures correctly when you brought them forward to the summary sheets. The clerk can assume that your judgment about manhours, quantities and material costs are correct. But be sure they verify everything else.

Here are the most common estimating mistakes. Eliminate these and you've cut errors by at least 80 percent.

The nine most common estimating mistakes

1) Errors in addition, subtraction, multiplication or division, especially decimal point placement. I suggest that you use a calculator, even for simple math that you could do in your head.

2) Errors in copying subtotals from one sheet to another.

3) Omission of an important material, labor, equipment, or overhead item. Use a checklist to call your attention to anything you've left out. In Chapter 4 you'll find good checklists for residential, commercial and industrial construction. Never consider an estimate complete until you've reviewed every word of the specifications and found every last dollar of construction cost.

4) Errors in estimating the manhours required to complete a task. I've devoted an entire chapter to recording costs. That's Chapter 8. Keep good records on crew productivity and you'll make better manhour estimates.

5) Obvious errors that can't possibly be right. Don't write down a figure that doesn't make sense. If it seems wrong, it probably is.

6) Using non-standard abbreviations. Abbreviations invented on the spot spell trouble. Here's an example. Several years ago I had a job that required 200 feet of base shoe. In the trim section of my lumber list, I wrote "200′ ¾″ B.S." The lumber yard must have had a new hand on the order desk because they sent 200 board feet of ¾-inch beveled siding.

7) Errors multiplying or dividing fractions. Convert fractions to decimals any time you're going to multiply or divide. I know that multiplying and dividing fractions is only sixth grade arithmetic. Anyone can do it. But not with consistent accuracy. A score of 90 percent probably got you an A grade in elementary school. But it's not good enough in an estimator's office. Convert fractions to the decimal equivalent and use a calculator.

8) Errors in units of measure. Be careful to distinguish between linear measure, square measure and cubic measure. For example, use one tick mark (′) for feet and two tick marks (″) for inches of linear measure, never for square measure or volume. Twenty-five square feet should never be written 25 sq.′, but always 25 SF or 25 square feet.

9) Illegible estimates. Write as clearly as you can. Make your estimates understandable so others can check them easily. Don't try to cram too much information into too little space. Paper is cheap. Mistakes are expensive. So use lots of paper. Leave plenty of space and blank lines on your take-off and summary sheets.

As I said earlier, this chapter was intended to fill any "holes" in the background information most estimators use. That does it for estimating background. In the next chapter I'll explain the steps I go through to make an estimate.

Chapter 4
The Estimating Process

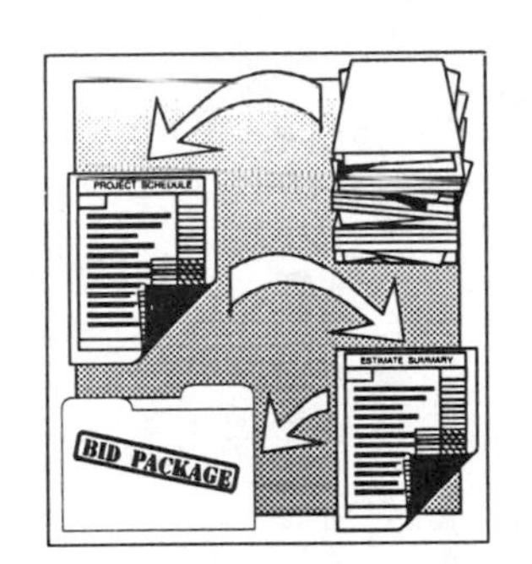

What I've learned from conversations with the hundreds of estimators I've talked to is that most experienced estimators use about the same procedures to compile estimates. They don't always do it in the same order, and they don't follow every step on every job. But the similarities far outweigh the differences.

I've identified fifteen distinct steps most estimators follow when they bid jobs. Good estimators use them for every bid, even the simple ones in which they might be tempted to "wing it." These fifteen steps are the focus of this chapter. Later in this manual we'll examine several of these steps in much more detail.

Detailed Estimating Steps

I believe the fifteen steps listed are essential for accurate construction estimates. And probably, if your estimates are always detailed, accurate, reliable estimates, the procedure you're following now includes most of these steps. Review my fifteen. There might be some tricks you don't know about.

Always follow these steps in sequence. Each step builds on the previous one. I suggest that you do them in the order listed in Figure 4-1.

Some of the steps are very straightforward. We'll discuss them very briefly in the paragraphs that follow. Others are more troublesome. I'll suggest some problem-solving techniques that should help you avoid the most common errors when you prepare construction estimates.

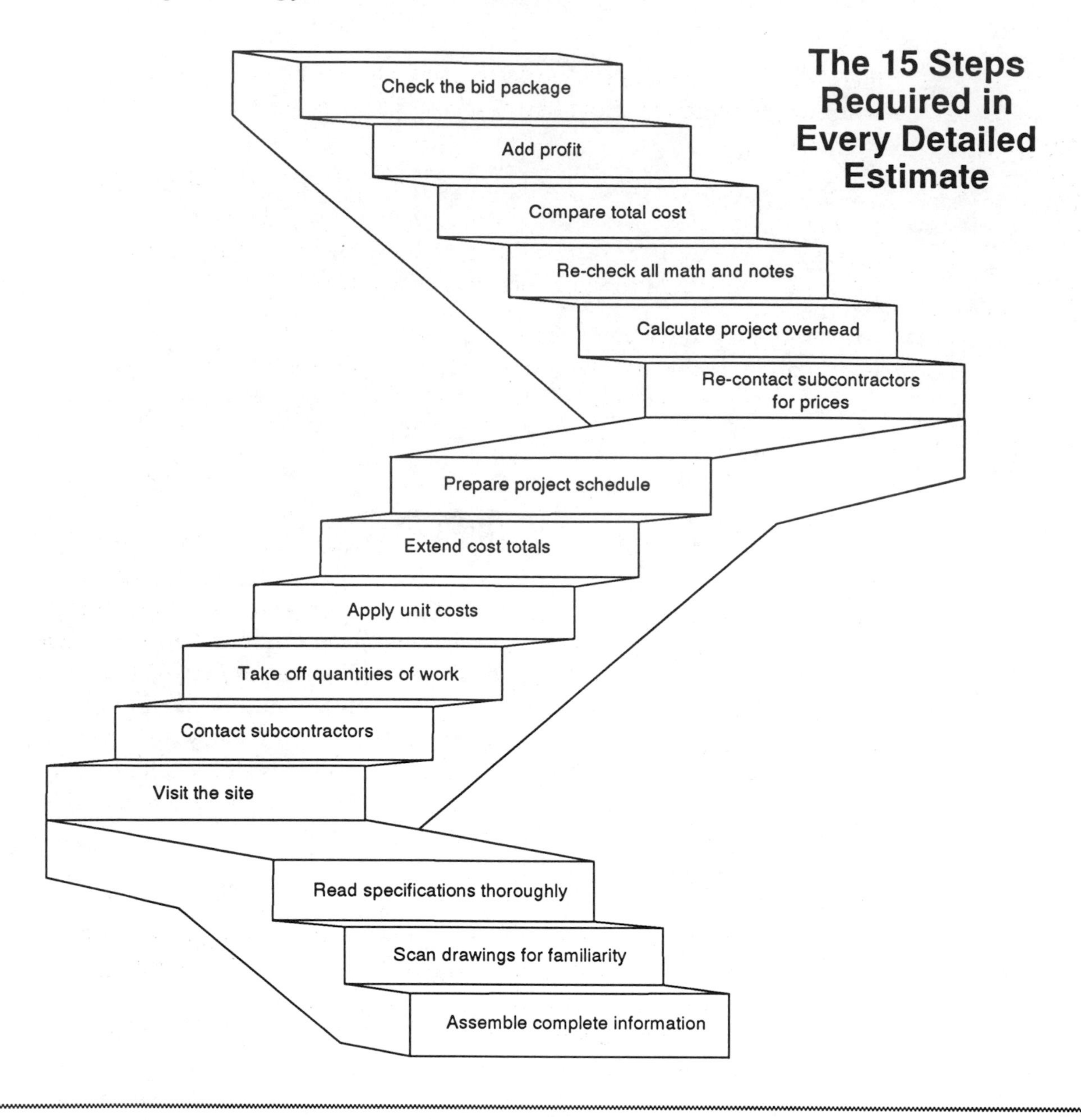

Figure 4-1 Fifteen steps to a detailed estimate

Step 1 - Assemble Complete Information

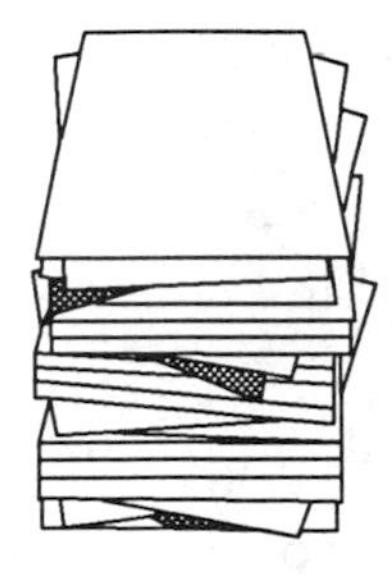

Before you begin any estimate, assemble all the documents that can affect construction costs. Along with an invitation to bid, you'll usually get a set of drawings, job specifications, and a sample contract. Be sure the documents you have are the most current available and are the same as other bidders will be using.

Talk with the person who requested the bid. That may be the architect, the engineer, the owner's representative, or the property owner. Ask if there's any information you haven't received yet. An architect may tell you there's an addendum in the works that will change the roof system. You don't want to find out about that after you've spent several hours doing the roof take-off.

Know who your competitors are

When you make this call, try to find out who's bidding against you. You probably know of some lowball contractors you'd rather not bid against. If several notorious cutthroat artists are already fighting over this bone, save your time for more productive work.

Your prospective customer won't always be willing to give you the names of other bidders. But ask anyway. On government work there's usually a published bidder's list. Take the time to find out who's on that list before you spend any time bidding the job.

Estimating takes time and costs money. The typical cost of preparing a detailed estimate is ¼ to ½ percent of the cost of the work. Suppose a potential client asks me to bid on a custom house valued at about $200,000. My cost for doing a full take-off will be between $500 (¼ of 1 percent) and $1,000 (½ of 1 percent). Most of that represents my time as estimator and the cost of paying someone to check my work. Other expenses include time to visit the site and call suppliers, and the clerical expense for preparing the bid itself.

Assuming you're satisfied that there's a profit to be made on this job, go to Step 2.

Step 2 - Scan the Drawings

Do you really want this job?

As you begin to scan the plans, ask yourself this question: Do I really want this job? Avoid wasting time. Maybe bidding on this project is limited to contractors with bonding capacity far beyond what you can qualify for. Maybe it's a job that requires special skills you don't have. Maybe it requires work by types of subcontractors that you've never dealt with before. Think about these things in the first five minutes you spend looking over the plans, not after you've put in a half day doing the take-off.

The specifications you're asked to make a bid on may be no more than some likes and dislikes expressed by the owner. On the other extreme, a major government project may be several bound volumes. Take a few minutes to study the plans and get thoroughly familiar with the project. I explained the procedure in Chapter 3.

Ask yourself these questions

As you consider whether you really want to bid this job, look carefully at the drawings and answer questions like these:

1) What type of construction is it? Am I equipped to do this kind of work?

2) Is my company usually competitive on this kind of job?

3) Is there anything unusual about the drawings?

Suppose I note that the structural drawings require a pre-engineered Butler-type frame. If two of my biggest competitors are franchised Butler installers and I'm not, why waste my time bidding this job?

Assuming the job fits your company, continue to Step 3.

Step 3 - Study the Specifications Carefully

The first two steps (collecting documents and scanning the drawings) are pretty straightforward and don't take much time. This step takes longer, and it's much more important: Read the specifications thoroughly.

On a large commercial project, the specs may be a bound volume of 100 pages or more. On a job like this, keep a pad and pencil next to you while you read the specs. Make notes about anything you run across that's unexpected or unusual.

Most of what you find in the specs will describe standard construction methods. If you're bidding a frame structure, you might run across something like this for the outside walls:

All exterior frame walls to be constructed of construction grade 2 x 4s placed 16" OC, with a single plate bottom and a double plate top, with two 16d fasteners at each location.

You probably won't bother to make notes on that. That's the way your carpenters have framed walls for many years. But if the specifications call for treated lumber for all exterior walls, you should make a note of that, and refer back to your notes when you're preparing the bid. Your notes should reflect the things you don't normally do. If the job is similar to most of the work you handle, you many need very few notes.

Read the specs for work you'll sub out

Even if you usually subcontract most specialty work, read the specs for those specialties. Otherwise you're buying trouble and maybe even a lawsuit. Whether or not your own crews do the work, you're responsible for the entire job. Your subs will read the specs, of course. But do them a favor. Call their attention to anything that's unusual in the plans or specs.

For example, suppose you notice in the specs that stainless steel reinforcement is required in the concrete. The drawings show the rebar but don't say anything about stainless steel. Make sure your concrete subcontractor is aware of the requirement and includes stainless steel reinforcement in his bid. The difference isn't just 20 or 30 percent more than standard rebar — it's several times more. And don't say, "If he doesn't notice it, it's his funeral." It could be yours, too.

What if the plans show regular ASTM A615 Grade A60 bars and the specs require stainless steel bars? Of course, I'd first call the architect or owner and request a clarification. But the general rule is that where there's an apparent conflict between the plans and the specifications, it's the specs a judge will turn to when a case appears in his court.

General contractors are concerned with all pages of the specs, of course. But even specialty contractors should read more than the specs for their section. For instance, if you're a masonry subcontractor, you'll concentrate on Division 4 (Masonry) of the specifications if the specs follow the CSI format. But be sure you read Division 1, General Conditions, also. These general conditions apply to every contractor on the job.

Step 4 - Visit the Site

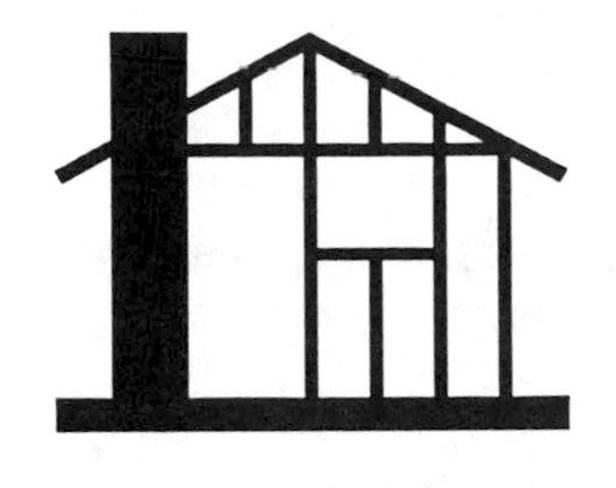

Don't omit this step! Always check out the site before you bid any significant job. Of course, it's possible to bid from just the drawings and specifications. It's also possible to drive through red stoplights. But it's risky. Treat this step like cheap insurance. If you discover just one major cost item that you may have overlooked, such as a swamp, it'll more than pay for a full year of site visits.

Look for obstructions

Many construction sites have physical barriers or conditions that make work more difficult. Problems like that never appear on an architect's drawings. All you have to do is drive by and you'll recognize them. Notice the steep driveway that's sure to be a mud bath after the first spring rain. Check out the pine tree that's going to obstruct access to the rear of the lot. Be sure there's enough room to store tools and materials on site. Make notes about what type of protection they'll need for overnight storage.

There are some communities where you wouldn't hesitate to leave a cement mixer chained at the site and unopened sacks of mortar stacked under a tarp. But in some neighborhoods, you'd better plan to take everything movable back to your yard every night. If you have to do that, and then drag it all back to the site again the next morning, you'll reduce your work day by at least one hour. Think about that, it'll cost you plenty. But it may cost you more to replace all the items left behind if they're stolen.

And how about availability of supplies? If there's no building material dealer in the community, someone is going to have to chase down tools or materials needed at the last minute to complete the work. It hurts to lose an hour of a carpenter's time while he drives 20 miles to replace a broken drill bit. If there isn't a hardware store nearby, you'll have to spend more time anticipating the supplies your work crews will need daily.

Who has to catch all these cost variables? The estimator, of course. And the best way to catch them is to visit the site.

Keep a checklist of unexpected costs you've encountered in the past. Take that list with you when you visit a job site. Figure 4-2 is my job site checklist. It could save you thousands of dollars — but only if you use it.

[ ]	Access for utilities	[ ]	Protection during construction
[ ]	Accessibility to the site	[ ]	Removing utilities
[ ]	Barricades	[ ]	Security
[ ]	Concrete cutting	[ ]	Street closing fee
[ ]	Debris removal	[ ]	Street repair bond
[ ]	Dust partition	[ ]	Street repair fee
[ ]	Job phone	[ ]	Supplies available nearby
[ ]	Job shack	[ ]	Temporary utilities
[ ]	Job signs	[ ]	Transportation of materials and equipment
[ ]	Material storage facilities or space	[ ]	Travel time
[ ]	Parking	[ ]	Vegetation, soil conditions and topography
[ ]	Protecting adjoining property		

Figure 4-2 Job site checklist

Step 5 - Contact Your Subcontractors

If you've been in business in the same town for several years, you probably have a list of subcontractors you use regularly. These are people you're comfortable working with, who know what you expect of them, and who are willing and able to do the job for you. But what if you're bidding a job in another city? Or what if your favorite drywall contractor retires and all his hangers and finishers go to work for the brother-in-law you haven't spoken to in years? Who do you call?

Don't look in the yellow pages

My favorite starting place for subcontractor referrals is my local building material dealer. Suppliers who specialize in a particular trade can provide the best referral for that trade. Don't go to the yellow pages. Anyone can advertise there. In fact, you may find ads in the yellow pages for guys who've had their licenses yanked and been out of business for a year.

If you need a good concrete contractor, ask the major concrete material suppliers for their recommendations. Talk to as many as you can and have them each recommend a few contractors for the size job you're figuring. Do this and you're sure to get a reliable subcontractor. The last person a supplier is going to recommend is a guy who doesn't pay his bills on time.

After talking to several suppliers, call the subcontractor who's mentioned by more than one of your sources. If you ask two concrete suppliers for referrals and three of the six recommendations you get are on both lists, those are the three to call. All three are probably smart, competitive managers. They don't automatically buy all their materials from the same place all the time. They shop around to get the best prices and they are well known. They save money where they can, and they can probably pass some of their savings on to you!

Talk to the superintendents in the area

If you're bidding a job in a new community, talk to several job superintendents in the vicinity. Look for a job that's similar to the one you're bidding. See whose trucks are parked at the job. Pay attention to the appearance of the job site. You'll decide pretty quickly if that crew should be working on your job.

Look for a job where everything is cleaned up right after the workers finish. See if materials are stockpiled neatly and conveniently. A job boss who has materials delivered to a spot far from where they are to be installed probably isn't a good planner — and won't be competitive.

Talk to general contractors or superintendents on the job. Say, "Hi, I see you have Jones Electrical working for you. How are they doing?" If there's something wrong with their work, the super will probably give you an earful. He might say something like, "Well, they're cheap enough — our office uses them on half our jobs. But they're 2½ weeks behind schedule, and they still don't have enough people down here."

If a local contractor who works with that sub regularly can't get them to perform, how are you going to do it from 100 miles away? That's the kind of subcontractor headache to avoid.

Inspectors and supervisors are another good source of referrals. The fire marshall knows the most reliable fire-protection specialists.

Build a file of available subcontractors for all the trades you use regularly. Supervise them closely so you get to know their abilities. Always contact more than one sub for every trade on every job. More contacts mean lower prices. And consider getting bids from subcontractors for work your own crews usually do. You might find someone out there who can do the work cheaper than you can.

The correct way to request a bid from subcontractor

Here's the right way to put a job out for bid to a subcontractor:

- Furnish the sub with a written request for bid, complete with specs and drawings.
- Be very specific about the work that's to be done. Don't rely on the sub to get the information from the specs.
- Tell the sub when you'll need the work done and ask for a written work and price schedule. Ask if their workload will permit them to meet your schedule.
- Never depend on a new sub to perform until you've checked that sub's references.

Step 6 - Prepare the Take-off Sheet

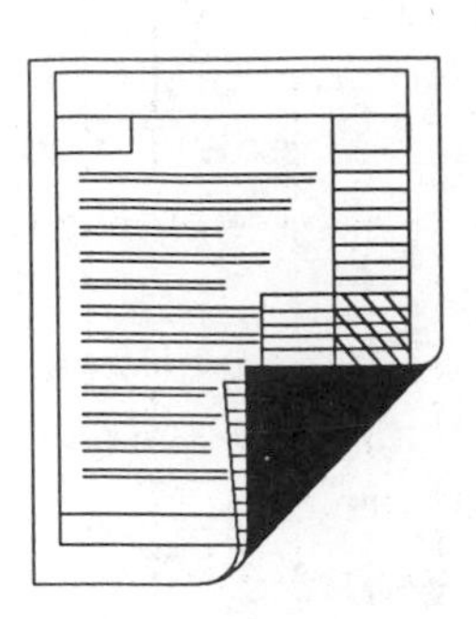

Here's where the real work of preparing the estimate begins. Figure 4-3 is my take-off sheet, reduced about a third so it'll fit in this book. Other supplies you'll need include a scale, colored and regular pencils, and a calculator. On the take-off sheet you'll list every cost item required to finish the project.

Follow the same order when you list costs on your take-off sheet as you will need them when you build the job. That helps prevent omissions, because you can visualize each step as it's completed. As I explained in the last chapter, missing an important cost item is the estimator's cardinal sin. The estimated cost of what you missed is always zero dollars. That's the worst possible underestimate. You'll reduce omissions by estimating costs in the same order that you do the work.

If I'm the prime contractor on a job, I know that the site requires clearing, grading and preparation, that underground utilities will be installed next, then the foundation laid, and finally the superstructure completed. If I'm the concrete sub, I have to put in the pile caps, then the foundation wall, then pour the slab. In each case, if I do the take-off in the order I'll do the job, I'm less likely to miss something, like the pile caps, for instance.

If you use a spreadsheet program such as Lotus 1-2-3 or Microsoft Excel, turn Figure 4-3 into a template for your estimates. If you don't use a computer, use Figure 4-3 as your summary sheet to combine figures from all sections of the estimate.

Use colored pencil to mark your drawings

Mark your drawings - Use colored pencils to make a mark on the drawing as you list each item on your quantity take-off sheet. Mark every note, every line and every detail as you write it on your take-off form. That makes it easier to see at a glance anything you've missed. When you review the plans, you'll see immediately if, say, a little stub wall isn't marked. It was probably missed. If you've marked it, you can be pretty sure it's listed on your take-off sheet.

Chapters 5 and 6 have lots more information on doing the take-off, including sample estimates.

List your BBD - *BBD* is an abbreviation for *basic building data*. This is the information or numbers you'll use over and over again on the estimate. For example, the floor area, the wall area and the building perimeter are all BBD. In order not to have to measure and re-measure these dimensions every time you need them, make a list of BBD.

You'll use these figures many times. For instance, once you calculate the total square feet of floor, you'll use the same number for the slab area, the floor covering and the suspended ceilings. You don't have to re-measure every time. Just refer to the BBD form.

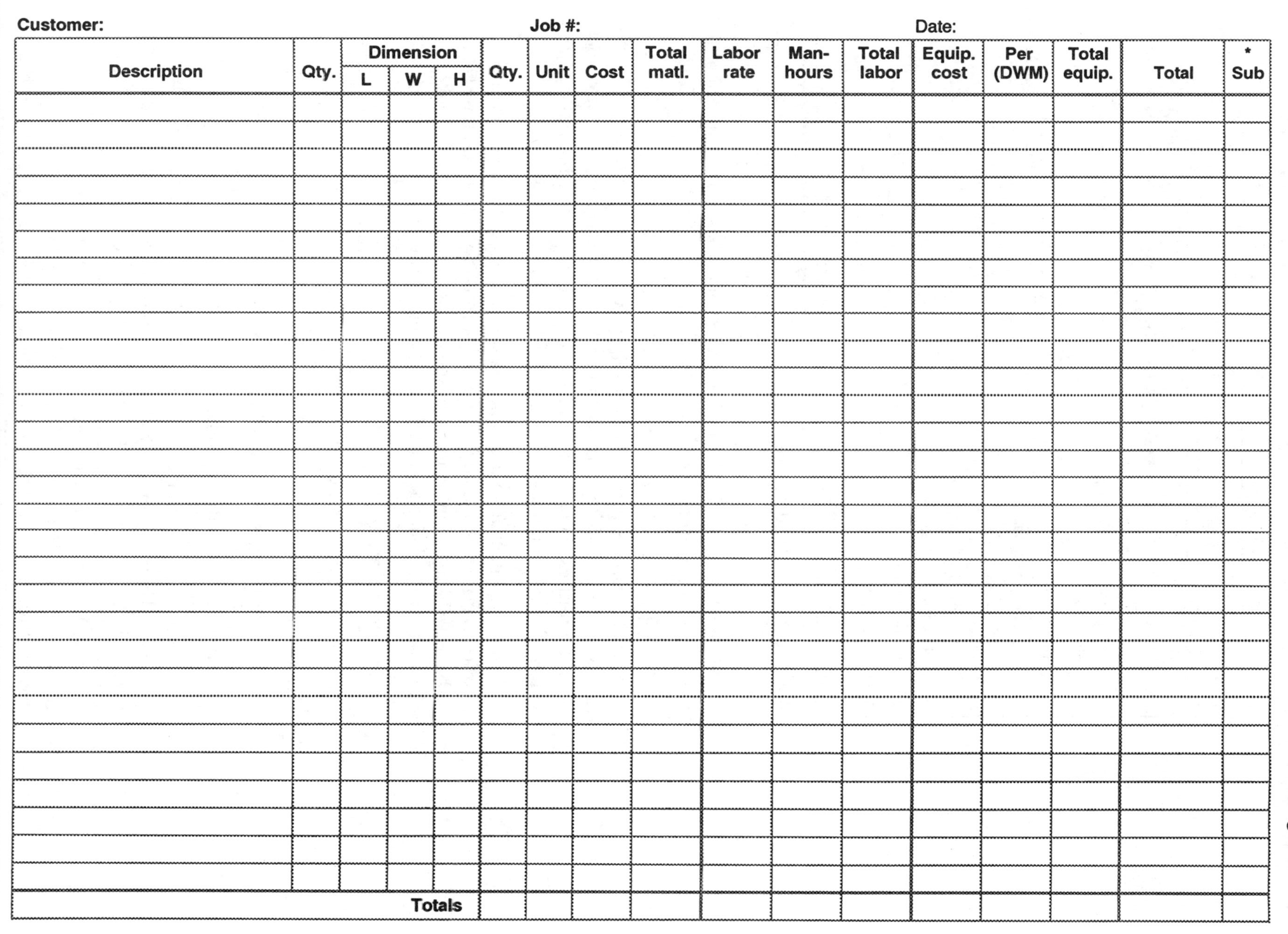

Customer: Job #: Date:

Description	Qty.	Dimension			Qty.	Unit	Cost	Total matl.	Labor rate	Man-hours	Total labor	Equip. cost	Per (DWM)	Total equip.	Total	* Sub
		L	W	H												
Totals																

Figure 4-3 A sample take-off sheet

Customer ____________ Job# ________ Date ________

Basic Building Data

Lot size	________	SF
Paved area	________	SF
Building footprint	________	SF
Building perimeter	________	Ft
Excavation depth	________	Ft
Excavation volume	________	CY
Number of floors	________	Ea
Total floor space	________	SF
Total wall space	________	SF
Total air-conditioned space	________	CY

Figure 4-4 Basic building data

Figure 4-4 is the BBD form I use. Copy it to the top of your take-off sheet, or make it a separate page.

Sometimes you'll adjust the BBD numbers. For instance, you subtract the square feet of hard flooring (wood, vinyl, or tile) from the total square feet of floor to determine how many square feet of carpet the job requires. Then you divide by nine to calculate the total square yards of carpet and pad.

I recommend that your detailed cost estimates show every significant cost item. For example, you wouldn't list just "5100 SF of 8-inch elevated slab." That's an assembly of costs and trades, not the smallest detail. The elevated slab includes at least the following cost items:

3rd floor requires:

124 CY 3000 psi concrete

7300 Lbs #6 bar

1403 Ft edge form or bulkhead form

5100 SF steel trowel finishing

Of course, those aren't all the costs that could be included in this assembly. There are probably many more. And your job as an estimator is to find them.

Why bother estimating every last cubic yard of concrete and board foot of form lumber? Because that's the only way to compile accurate construction cost estimates. It's not enough to list "5100 SF of reinforced slab" on your estimate form. The cost for 5100 square feet of slab can vary too much from job to job, even if the total square footage remains the same.

For example, suppose the floor slab has an irregular shape. There's more bulkhead or edge form on this job than there would be if the floor was a simple rectangle. You've got to break out the details for an accurate estimate: formwork, concrete, rebar, finishing. Then estimate the labor, material and equipment cost for each.

Step 7 - Apply Unit Costs

When you've listed every cost item, begin adding prices to your take-off sheet. Material costs aren't usually a problem. A trainee estimator with 30 days' experience can fill in most of those. Just call your suppliers, tell them what you need and how much, and where they'll have to deliver it. Then write in the lowest price on your take-off form.

Estimating labor costs isn't quite as simple. For instance, to estimate the cost to place concrete, you have to know something about crew efficiency and job conditions. How long will it take to get the concrete to the form area, consolidate it, level it and finish it?

"Review records of similar projects you've completed to get an idea of the crew's productivity rate."

Pricing labor and material costs is an important part of every estimate. Chapters 5 and 6 cover pricing using the traditional pencil and paper methods. In Chapter 7 we'll give you some practice using a computer to price a job.

Review records of similar projects you've completed to get an idea of the crew's productivity rate. How many cabinets can your finish carpenter set in an hour? Will you need to adjust your normal labor rate for this particular job? What will slow the crew down? What will make the job go faster? You need cost records to answer questions like that. Chapter 8 explains a good way to keep a record of job costs so you'll have useful data available when you need it.

You'll need your firsthand knowledge of the site to complete this part of your estimate. Check your notes: Can the transit mix truck maneuver close enough to the form to chute the concrete into place? Or will you have to haul or pump the concrete? If you have 25 yards to place, is it situated so you can do it all at one time, or will you have to split the job over two days? The more you place at one time, the less it costs per unit. Here's where an estimator needs experience. Estimating isn't just measurements and arithmetic. It takes experience and judgment.

Step 8 - Extend and Total Costs

You may wonder why extending and totaling prices is a separate step. After all, you only have to multiply the quantity by the unit labor and material costs and then add up all the columns. But here's the unfortunate estimating fact of life: This is where most of the serious estimating mistakes occur. Whether you use a spreadsheet program or a calculator to extend prices and add columns, the risk of major error is substantial. Of course, if you use a computer for pricing, accumulating row and column totals should be automatic. See if you don't agree after you read Chapter 7.

Be consistent

Check your work - The best way to minimize math mistakes is to be consistent. Set up procedures that you follow every time. Show all your work to make checking easy. Have someone check for errors on every estimate — both the cost extensions and the column totals. No estimate is complete until someone has taken responsibility for the math by initialing the line, "Checked by _____ ."

I like to have an experienced estimator review my estimate. But that's a luxury that isn't always available to every estimator. You should still have someone check your figures. The checker doesn't have to be an experienced estimator, just someone who's good with a calculator and can work carefully. Have that person extend each item, total each page, and check the grand total. It sounds simple. I'm sure you agree that it's necessary. But it isn't always done.

There's another good reason to have a second person go over the figures. No matter how good someone is at math, the person who writes in the numbers is the one least likely to catch a mistake. *Anyone* else will probably do a better job than you at finding your own math errors.

Don't tolerate sloppy work

Your estimates should be clean, clear and well-organized so they're easy to check. Don't tolerate sloppy work. Estimators have to write clearly. It's too easy to mistake a "7" for a "1." When numbers are copied from one sheet to another, have someone check the copy. It's also easy to skip a number or copy something to the wrong place.

Another common error is transposition, such as writing "9368" when you intend "9386." Again, get someone with an eye for detail to check the figures.

I also recommend that you save all your estimating notes and calculations. That's valuable backup detail if the estimate has to be revised or changed later. Keep your scratch papers, make lots of notes, and use a calculator that prints a tape. When you're finished extending and adding the figures, staple the calculator tapes to your working copies of the estimate. That creates an audit trail showing where your numbers came from and how they were computed. A $10,000 error discovered in time compensates for a lot of clerical pay.

A Sample Take-off

If you're like me, you learn best by doing. I've covered half of my fifteen estimating steps without giving you the opportunity to practice. Before we go any further, let's do a little take-off drill. In the next chapter you'll have a chance to get in lots more practice.

Suppose you have a set of structural drawings that show pile caps around the perimeter of the building in Figure 4-5. Figure 4-6 shows the piles and caps in plan and elevation view. Notice the driven timber piles and poured reinforced concrete caps. Figure 4-5 also shows some other details, but we'll consider only the PC-1 pile caps here.

You're the concrete estimator. When your crews move on to this job, the pile driving sub will have finished. Your job is to pour concrete caps on the piles. Let's do a take-off for the PC-1 caps.

3 categories for pile caps

You'll have three categories of costs for the pile caps.

First, you want to know how many cubic yards of concrete to buy and place.

Second, you need to figure how many square feet of forms you have to set.

Third, you need to know how many pounds of reinforcing steel you have to buy and place.

This is buried structural concrete, so you don't have to do any finishing. Your placement crew will float off the concrete. A machine-trowel finish isn't required, so you don't need to list trowel finishing on your take-off sheet.

Figure 4-7 is my take-off sheet. Notice that it has three columns for dimensions. You won't use all three columns for every item. If you're taking off squares, you'll use two. If you're measuring linear dimensions, you'll use only one column. You'll use all three if you're taking off volume.

Concrete Take-off

Figure 4-6 shows that the PC-1 caps measure 2′6″ by 5′6″. They're 2 feet deep. Always convert inches to hundredths of a foot to make your calculations easier. Figure 4-8 is your conversion chart. Notice at the top of the *Description* column in Figure 4-7 that I've listed the quantity of PC-1 caps (22) and the length, width and height of each cap.

I like to show my calculations on the take-off form. Notice my volume calculations under PC-1 caps. Some estimators like to make these calculations on a separate sheet. I have no objection to that. In any case,

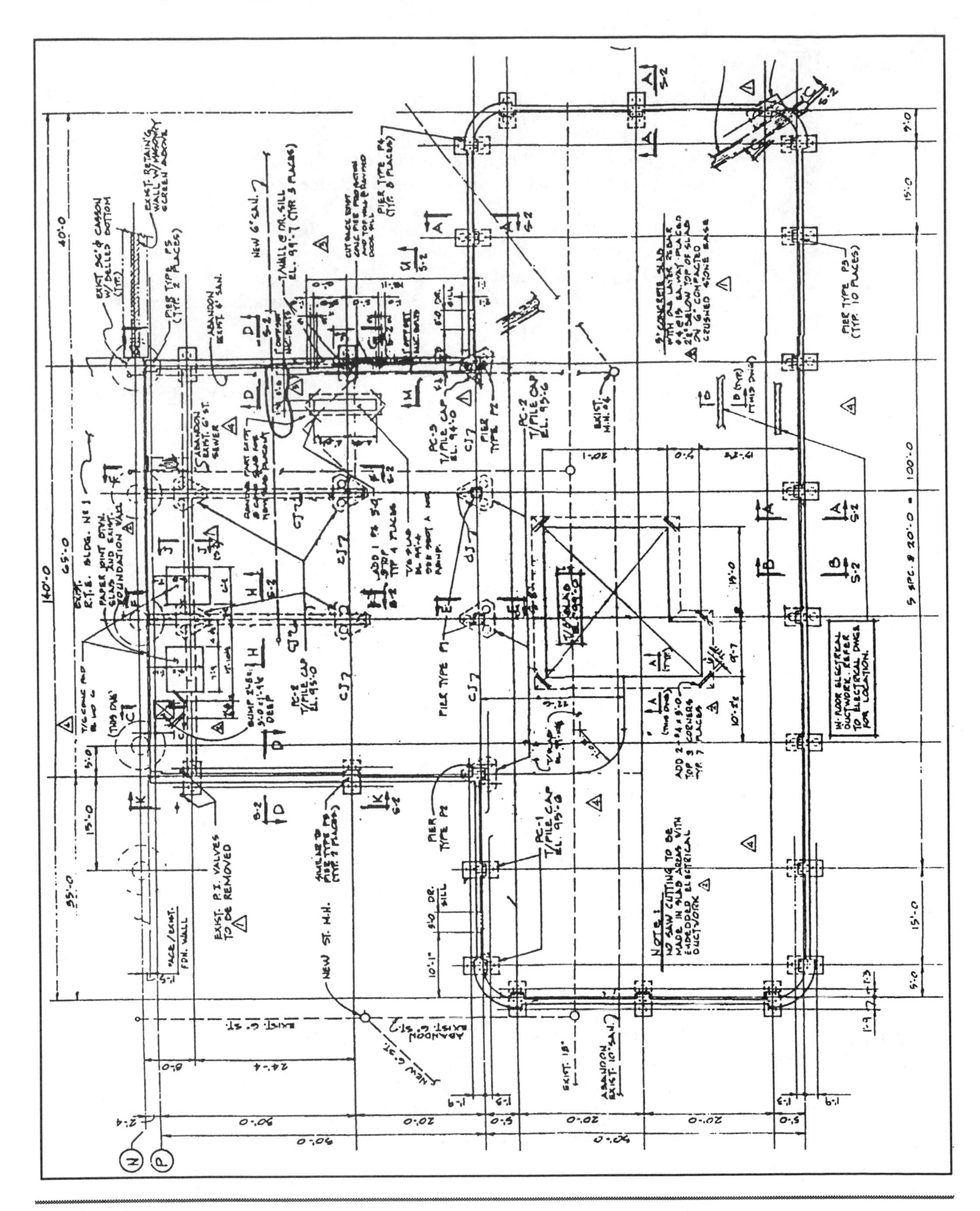

Figure 4-5 A working plan

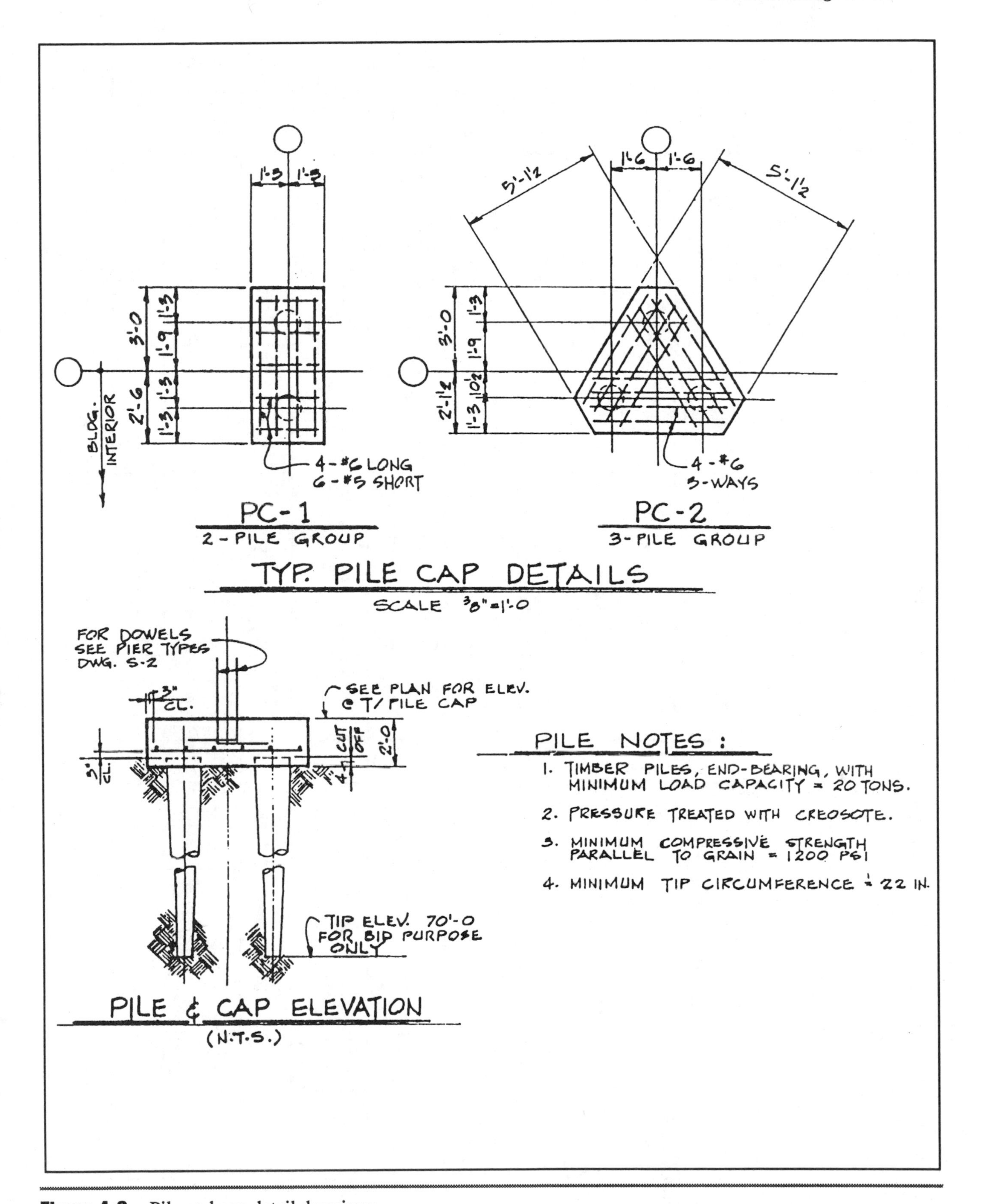

Figure 4-6 Pile and cap detail drawings

Customer: Job #: Date:

Description	Qty.	Dimension			Qty.	Unit	Cost	Total matl.	Labor rate	Man-hours	Total labor	Equip. cost	Per (DWM)	Total equip.	Total	* Sub
		L	W	H												
PC-1 CAPS	22	5.5	2.5	2												
4000 PSI CONC.																
2.5 x 5.5 x 2 = 27.5																
22 x 27.5 = 605																
+ 6% WASTE = 37																
642					24	CY										
FORMING	22	5.5	2.5	2												
PERIMETER																
2 x 2.5 = 5																
2 x 5.5 = 11																
16	16'			2'												
16 x 2 x 22					704	SF										
REBAR																
#6 @ 1.5 LB/LF																
4/CAP	88	5'														
5 x 88 = 440																
440 x 1.5					660	LB										
#5 @ 1 LB/LF																
6/CAP	132	2'			264	LB										
Totals																

Figure 4-7 Pile cap take-off sheet

Conversion chart: Inches to decimal feet

1	=	0.08	7	=	0.58
2	=	0.17	8	=	0.67
3	=	0.25	9	=	0.75
4	=	0.33	10	=	0.83
5	=	0.42	11	=	0.92
6	=	0.50	12	=	1.00

Figure 4-8 Inches to decimal conversion table

you'll need to multiply cap dimensions to find the volume for each cap. Then you'll multiply the volume of each cap by the number of caps to find the total volume.

Notice in Figure 4-7 that I've added 6 percent for waste. There's almost always some waste when you pour concrete. Round your calculations up to the next whole number. Then divide by 27 to find the concrete volume in cubic yards. Notice those calculations in the *Description* column of Figure 4-7.

Formwork Take-off

Assume you're going to use plywood forming for these caps. The dimensions you'll use are the same as the ones you used to figure the concrete volume. But here you're concerned with surface area, not volume. Multiply the perimeter of the caps (16 linear feet) by the cap depth to figure how many square feet of concrete forms are required.

Notice my forming calculations in the *Description* column of Figure 4-7. I've multiplied the length and the width by two because each cap has two side forms and two end forms. The cap perimeter is 16 feet. Then I multiplied the perimeter by the height. The answer (32 square feet) is the plywood needed to form each cap. Finally, I multiplied 32 square feet per cap by the number of caps, 22. The answer is 704 square feet of plywood.

Notice that I haven't included any plywood waste in the formwork estimate. There's very little waste when you apply sheathing and siding if you can use full sheets on a job. Here, there may be no waste. We'll cut the side forms and end forms for each cap out of an 8-foot sheet of plywood, split lengthwise. One 4- by 8-foot sheet forms each cap. If the formwork crew works accurately, only 704 square feet of plywood (22 sheets) will be needed.

Reinforcing steel take-off **-** Start by counting the steel bars in the PC-1 detail. As you count each bar (represented by a dashed line), put a check mark on the line with colored pencil. That shows you counted it. In this case, I count five short bars and four long bars.

But look again. The designer shows only five short lines to indicate #5 bars. Now look at the note below the drawing. The note clearly requires six #5 short bars. Remember that the notes always take precedence over the drawing, so be sure to check them. Here, we'll figure six short bars.

Reinforcing steel is usually ordered and estimated in pounds, but taken off the plans in linear feet. To convert lengths to pounds, we multiply by the weight per linear foot. Number 5 bars weigh about 1 pound per linear foot. Number 6 bars weigh about 1.5 pounds per linear foot.

Figure 4-6 shows four #6 rebars 5 feet long in each cap.

✏ 4 bars x 22 caps = 88 bars

✏ 88 bars x 5 feet = 440 feet

Number 6 rebar weighs about 1.5 pounds per linear foot. So we'll need 440 times 1.5, or 660 pounds of #6 rebar for PC-1.

I've used the same method to figure the #5 steel at 264 pounds. Notice there's no allowance for waste here either. Rebar comes in 20-foot lengths. The pieces we need are lengths which divide evenly into 20, so we don't have to allow for laps or waste.

Figure 4-7 shows the detail needed for the most accurate estimates. Cost manuals published by several companies show what claim to be current costs per cubic yard for pile caps poured in place and with reinforcing. But none will be as accurate as the figures you develop by measuring and figuring each cost item yourself.

Step 9 - Prepare the Project Schedule

Time is money

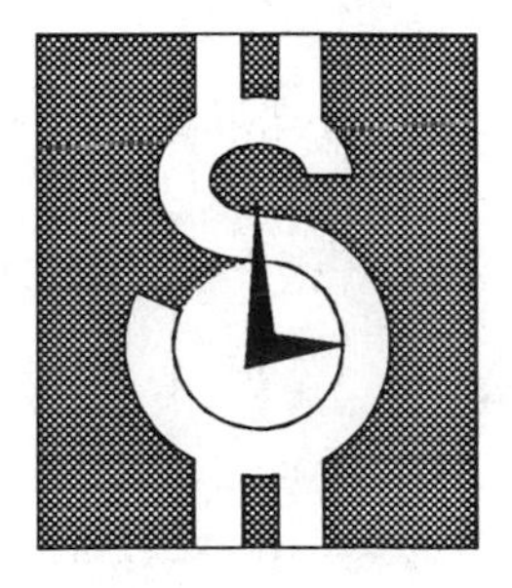

You've heard the saying, "Time is money." That's especially true in the construction business. That's why I feel that no cost estimate is complete until you've done some thinking about the project schedule. After you read the examples in this section, I expect that you'll agree.

I recommend that the senior estimator or project manager prepare a preliminary schedule for the entire project as part of the estimating process. This can be a simple listing of what has to happen (usually called an event) and how long it should take (the duration). Then put these events in order. What you have is a simple critical path chart. Later, once a contract is signed, the estimator or project manager will prepare a more detailed job schedule.

Show the sequence of events

Your preliminary schedule should show, in the correct order, the sequence of events needed to complete the work. For example, first, you have to clear the site and install underground utilities. Then you excavate and pour concrete. After that, the framing goes up, and so forth. Most of these events can't begin until other tasks are finished. That's the nature of construction. You can't hang wallboard until the wall framing is complete.

The schedule prepared as part of the estimating process is only an approximation. It doesn't have to include details like repairing access holes that aren't right or installing decorative hardware. You're not concerned at this point with scheduling crews and subcontractors. The project superintendent will handle details like that later. This schedule is intended primarily to predict the time needed to complete the project.

It isn't hard to lay out this kind of schedule. But it's an important step in the estimating and planning process. A good schedule will show you the most efficient, most cost-effective and most profitable way to do the work. Figure 4-9 is a job schedule for some work I estimated in Chicago recently.

Figure 4-9 shows that four weeks will be needed to pour and cure the foundation. That's based on my knowledge of the job and my crews. I know how many people I'll put on the job, how many feet of foundation they can form a day, and how much concrete they can pour and finish in a day.

But suppose there's a good reason to put this job on the fast track. Suppose I can beat most weather problems if the foundation is finished by a certain date. In that case, it may be to my advantage to consider what could be done to accelerate the completion date. The time to consider those alternatives is when you're estimating the job. I can decide to run the foundation work on two shifts, 16 hours a day, completing it in eight working days instead of 15, and figure the cost accordingly.

Of course, you don't have to schedule extra shifts to accelerate the work. You can also accomplish the same end by hiring extra workers to supplement your regular crew. Or, you might cut the time needed for plumbing or electrical work by having several plumbing or electrical subs working at the same time. There are many options available to you if you plan ahead.

Better scheduling means bigger profits - I once built an apartment project in a Chicago suburb with eight identical masonry structures. Masonry work fell on the critical path. That's the line drawn through the sequence of events that determines the final completion date. Instead of giving the masonry contract to the lowest bidder, I renegotiated the job with the two lowest bidders, each doing four buildings.

It would have been cheaper to have one mason do the entire job. In this case, renegotiating the masonry contracts cost me about $20,000 extra. But my goal wasn't to get the lowest cost on each bid. It was to maximize our return on investment and meet the buyer's construction deadline.

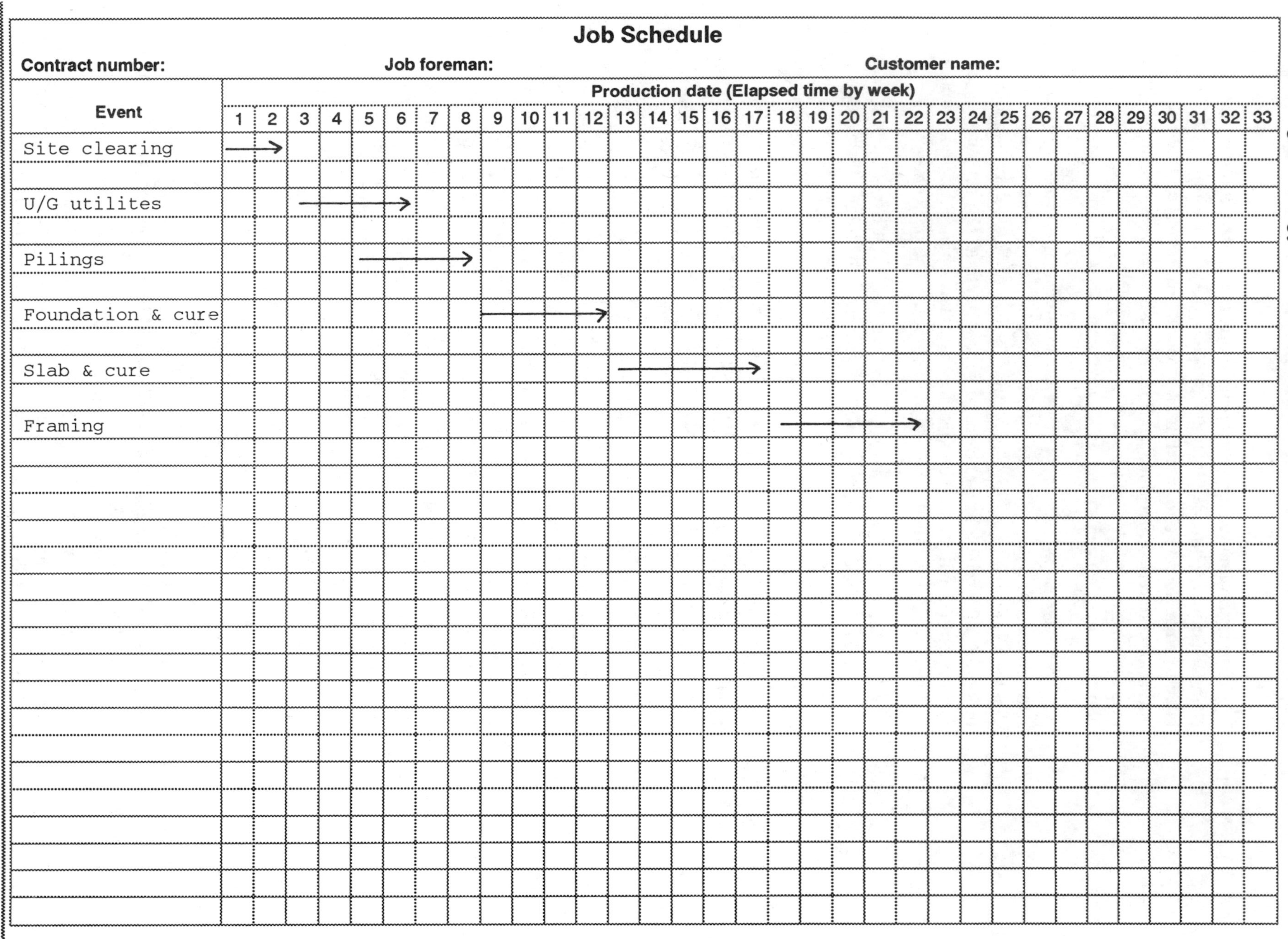

Figure 4-9 Estimating job schedule

Sometimes that means you spend more money on one item to reduce the cost of something else. On that apartment job, we saved money by using the fast track. Completing the masonry sooner kept other sub trades working almost constantly. That reduced scheduling problems and nearly eliminated unexpected delays.

There are other ways besides hiring extra subs, increasing workers or paying overtime to accelerate a project. Consider pre-assembly off site. It's possible, for example, to build 2′ x 2′6″ x 5′6″ reinforced pile caps as precast units instead of pouring them in place. Precast caps usually cost more. But if your overhead on a job is running $5,000 a day, shortening the schedule by a couple of days or a week can save plenty.

"Pre-assembly doesn't usually save much money, but it does save time."

Here's another example of how estimating and scheduling interact. A few years ago I had to estimate an 11-story concrete frame building in Marquette, Michigan. The plans assumed the building would be poured in place. All the other bidders quoted a two-year buildout. It's too cold and windy from November to March to pour concrete in northern Michigan on the shore of Lake Superior. The other bidders assumed that work on the project would have to be spread over two years. Initially, we planned our bid like the others. We would work seven months, pull off the project for five, then start again when it was warm enough to pour concrete again.

Then, I had an idea. My company invested an extra $50,000 in that job at the estimating stage to hire a structural engineering firm. They redesigned the job using precast members that could be assembled much more quickly. We finished the job in seven months using precast concrete. Doing that saved us several times the $50,000 we invested in consultants.

Pre-assembly doesn't usually save much money, but it does save time. When time is worth hundreds or thousands of dollars a day, think about where you might be able to save time when you prepare estimates. That's why scheduling has to be part of the estimating process. It forces you to think through the best, most cost-efficient way to complete a project.

Step 10 - Re-contact Subcontractors

Once you've roughed out a job schedule, it's time to begin collecting subcontractor bids. This is a major headache in most estimating offices. Nearly all subs prefer to wait until the last minute to submit their prices. They've had the drawings for two weeks and probably finished the estimate three days ago. But they won't call in their price until 20 minutes before you have to submit your bid. Now you have only minutes to check subcontractor bids, select the one you want and then plug the figures into your estimate.

Why do subs force general contractors to work under so much pressure? To prevent bid shopping. In the construction industry, many general contractors have a reputation for shopping their bids — taking competitive bids from many subs, then giving the job to the sub of their choice at a negotiated price.

If I were a sub, my best defense against bid shopping would be to hold off my bid until the very last minute. That way, there's no time for negotiating. My competitors wouldn't find out how much I bid until it's too late to change their bid. Everything considered, "crisis" bidding is probably good for subcontractors. Unfortunately, it creates panic among general contractors.

Last-minute bidding makes errors easy and even likely. When those last few minutes are ticking away, you need lots of experience and good judgment to make smart decisions. Here's an example.

Suppose I'm the chief estimator for a general contractor. I've just received a bid from a masonry contractor. We've worked with this sub on several small jobs in the past. I didn't invite him to bid on the job I'm figuring this week because it's bigger than he's handled for me before, about $90,000 compared to $10,000 or $12,000 in the past. But he found out I was bidding the job, got a copy of the plans and specs and called me with his price 20 minutes before I'm supposed to submit my bid.

This masonry sub's price is $58,000. I received two other bids earlier in the day, one for $93,000 and the other for $96,000. The exceptionally low bid raises a number of questions. If I didn't know this subcontractor and his work, I'd probably ignore the bid. There's no advantage in accepting a bid that's a mistake. But because he's been reliable in the past, I have to make some quick decisions!

I know this masonry sub has been in business less than two years. The jobs he handled for me in the past went all right. He finished on time. His workmanship was good. I haven't heard any complaints from suppliers about his bill-paying record. But here he is with a price that's nearly 40 percent below the competition.

I'm sure at least four other general contractors are bidding this work. I'm almost as sure that this lowball masonry sub called all four with his $58,000 price. I have to submit my price in just a few minutes. What should I do?

I have a couple of choices. Neither is ideal.

I have reservations about the bid, but I could use it anyway. If it turns out that the sub made an error, I run the risk of having to make the job right after he defaults.

I can use the next lowest bidder's price, $93,000. But my competition may use the $58,000 price and shut me out of the job.

Always have specs and take-off sheet handy

To avoid this situation, I recommend that whenever you take a phone bid from a subcontractor, have the specs and your take-off sheet handy. Ask the bidder for the quantity of two or three key materials used in the bid — in this case 8″ x 8″ x 16″ block. Ask for that material quantity before you even discuss the price. That's not "shopping" the bid. It's just verifying that the bid received covers the job you're estimating. If the other bidders have included twice as much block in the job, something is very wrong somewhere.

I also recommend that you prepare a quantity take-off for the whole job, including the trades done by subcontractors. If your sub's quantities are similar to your quantities and similar to quantities estimated by other subs, you're probably comparing apples to apples.

Be sure you have clearly defined the scope of work for every bidding subcontractor. That's especially important when you accept an unsolicited bid. The bidding subcontractors must be aware of everything that will go into the job. There's no advantage to winning bids that will bankrupt your subcontractors.

Request a performance bond

And finally, the best way to protect yourself from unreliable subcontractors is by requesting a performance bond. A performance bond is a guarantee by the bonding company that the work will be completed for the amount of the contract even if the subcontractor can't continue. Many owners require performance bonds from their prime contractors. Get the same from your subs. Your bonding company probably requires it for certain subcontractor trades.

I'll admit that getting bonded is a nuisance. It's like applying for a loan. Bonds increase your overhead and add to your cost. Plus, they contribute nothing to job quality. But performance bonds offer a unique form of security that's the best guarantee of reliable performance.

Reliable performance probably isn't a concern when the price range between high and low bidders is only a few percent and all subcontractors are well known to you. But where the range between high and low bid is nearly 40 percent, you need some protection.

Subs that have been in business for only a few years and are bidding jobs larger than usual seldom qualify for performance bonds. Bonding companies like to see lots of financial strength (money in the bank) and a selection of similar projects completed. Very few new construction companies can qualify.

So, what do you do about the $58,000 masonry bid? Here's my suggestion. I hope you can use it someday.

Before I accept any exceptionally low bid, I describe the problem to the low bidder and ask for an explanation. Maybe they're bidding low to get their feet wet in larger jobs. Maybe they're planning to absorb some labor costs to get the work. Maybe this is the kind of work they want to specialize in. Maybe competition is moving into the area and they want to shut the door. Or maybe they need fill-in work between two jobs.

At least advise the contractor of where their bid stands and get an explanation. Maybe the contractor knows he's bidding low. That's probably O.K. But contractors who submit lowball bids without realizing their mistake don't stay in business very long.

Subs to avoid - The sub bid I won't use comes from the specialty contractor who says, "Gee, I had no idea we'd be that far off. Let me call you back in five minutes."

We've already got his material quantities, so he can't change those. When he calls back, he says, "You're right. We were awfully tight on that one. We'll have to go up to $85,000. We probably couldn't have done it..."

He's admitting that he's in over his head. He doesn't know what it's really going to cost to complete the work. And so I'm not going to use his $58,000 price. And I'm not going to use his $85,000 price either. I'm going write in the $93,000 price submitted by a more reliable contractor.

What have I gained? The subcontractor who called me back with the new price is now calling all my competitors, raising his price to at least $80,000. That changes the odds in my favor considerably.

More problems with sub prices - Here's another headache. It's not quite as troublesome as the last one, but it's probably much more common. Suppose you've just finished an estimate. You've done your own take-off, visited the job site and collected prices from all bidders. You've requested material quantities from all four of the lowest bidders and they match perfectly. But there's still a problem.

The lowest bidder omitted one bid item. He says they don't install backdraft tension shields. You have a slightly higher price from another sub. They include the tension shields, but have substituted neoprene hanger joints for elastomeric friction couplers. You can't find anyone who can do the friction couplers alone. A third bid isn't clear about what type of joints or couplers are used. In short, the changes and exclusions in each bid keep them from being comparable.

Technology to the rescue - I think technology has created problems like these. Buildings have become high tech, just like computers and communications. But just as technology creates problems, so can it help you solve them. For instance, digitizers make it easy to do faster and more accurate take-offs. But, you still have to know what to take off. Otherwise you're just making mistakes twice as fast.

My favorite bidding aid is the fax machine. I've never seen an electronic gadget take hold in construction offices as fast as the fax. Use of computers and CAD (computer aided design) systems increased very rapidly during the 1980s. But today nearly every construction company doing commercial work needs a fax to stay in business. They're extremely valuable on bid day when rapid exchange of information is essential.

If you have any doubt about what's included and excluded in a particular bid, have the sub fax a copy of the scope of work they used to prepare the bid. You'll have it on your desk in 5 minutes. There shouldn't be any question about what the price covers. Even if the sub doesn't have a fax, most postal stores offer fax service at a modest cost.

Cover all the bases - My advice is to always insist on a written copy of your sub's scope of work. With a fax machine it isn't difficult to provide, and that way you don't leave anything to chance. Be sure you know exactly what's included in and excluded from each bid. That's a lesson most estimators have to learn the hard way. I did. Here's how it happened.

Early in my building career, I had two houses nearly ready for final checkout. Everything was finished. The furnace and central air conditioning units were properly installed. The thermostat was in the wall at the proper location. But there wasn't any low-voltage wiring between the thermostat and the heat exchanger units.

I called the HVAC contractor and asked when he was going to hook up the thermostat so we could run the air conditioner. He said, "As soon as the electrician runs the low-voltage wire."

So I called the electrician to complain that they didn't put the low-voltage wiring in. And he said, "We don't do low-voltage wiring. That's the responsibility of your HVAC guy." You can guess what the HVAC sub said when I called him back. I wound up paying extra to get the thermostats connected. Why? Because I didn't bother to ask my subs about the scope of their work. Back in 1974 when that happened, the extra wiring cost me only $35. No big deal. But the stakes could have been lots higher. A little oversight like that can cost you thousands. Be careful!

That kind of misunderstanding is common in the construction industry. I've heard subcontractors quibble over some pretty ridiculous things. Steel set in concrete seems to be a consistent problem. Who supplies the anchor bolts, the concrete sub or the framing sub? Who installs them? Does the framer send someone to work with the concrete crew to install the bolts? Or is the concrete contractor entirely responsible? Whose mistake is it if an anchor bolt isn't aligned correctly?

Even if anchor bolts aren't included in some subcontractor's bid, they're automatically included in *your* bid. That means you'll have to pay for them twice. Once to the guy you thought was going to do the job, and again to the guy you finally hire to do it.

Step 11 - Calculate Project Overhead

Don't confuse project overhead with company overhead. Project overhead costs are jobs costs — expenses that come from a particular job. Company overhead is the cost of running any business. It would remain about the same next week if you had no jobs at all. Project overhead continues only when jobs continue.

3 categories for project overhead

Project overhead is divided into three categories:

- Labor
- Equipment
- Other supporting costs

Project labor overhead includes supervision and other non-productive labor. Your superintendent is project overhead. So is the cleanup crew that does nothing but clean up. The watchman is project overhead. Time for these non-productive workers is charged to the entire project, not to any particular trade or task. Neither is it part of your company overhead.

Project equipment overhead isn't the same as equipment job expense. For example, if you do your own grading, you'll charge the cost of that D-4 dozer to excavation expense on each individual job. That's not overhead.

Project equipment overhead costs are the Division 1, General Conditions items. Examples are your office or storage trailer and the superintendent's pickup truck, including its operating expenses.

Other project overhead costs include project management, part of your own salary as estimator or supervisor, temporary utilities, sanitary facilities and trash removal.

It's easy to miss project overhead costs when you're estimating a job. Nowhere in the plans or specs will you ever see a note: "Project Overhead" and a list of your overhead costs. You have to dig these costs out for yourself. My suggestion is that you develop a good checklist and use it on every estimate. For residential work, I recommend you start with Figure 4-10. Add to it as required. Eventually you'll develop a very complete list of all the overhead items you're likely to encounter.

Of course, you won't have all the project overhead expenses listed on Figure 4-10 on every job. But go through the list before you turn in any estimate. Use it as a reminder system. Make a decision on every line. If it's a likely expense, write down some figure in the cost column. It doesn't have to be exact, just close. If you put in nothing at all, you're saying it's free.

Overhead Checklist

Project Overhead

- _____ [ ] Barricades
- _____ [ ] Bid bond
- _____ [ ] Builder's risk insurance
- _____ [ ] Building permit fee
- _____ [ ] Business license
- _____ [ ] Cleaning floor
- _____ [ ] Cleaning glass
- _____ [ ] Cleanup
- _____ [ ] Completion bond
- _____ [ ] Debris removal
- _____ [ ] Design fee
- _____ [ ] Equipment floater insurance
- _____ [ ] Equipment rental
- _____ [ ] Estimating fee
- _____ [ ] Expendable tools
- _____ [ ] Field supplies
- _____ [ ] Job phone
- _____ [ ] Job shanty
- _____ [ ] Job signs
- _____ [ ] Liability insurance
- _____ [ ] Local business license
- _____ [ ] Maintenance bond
- _____ [ ] Patching after subcontractors
- _____ [ ] Payment bond
- _____ [ ] Plan checking fee
- _____ [ ] Plan cost
- _____ [ ] Protecting adjoining property
- _____ [ ] Protection during construction
- _____ [ ] Removing utilities
- _____ [ ] Repairing damage
- _____ [ ] Sales commision
- _____ [ ] Sales tax
- _____ [ ] Sewer connection fee
- _____ [ ] State contractor's license
- _____ [ ] Street closing fee
- _____ [ ] Street repair bond
- _____ [ ] Supervision
- _____ [ ] Survey
- _____ [ ] Temporary electrical
- _____ [ ] Temporary fencing
- _____ [ ] Temporary heating
- _____ [ ] Temporary lighting
- _____ [ ] Temporary toilets
- _____ [ ] Temporary water
- _____ [ ] Transportation of equipment
- _____ [ ] Travel expense
- _____ [ ] Watchman
- _____ [ ] Water meter fee
- _____ [ ] Waxing floors

Company Overhead

- _____ [ ] Accounting
- _____ [ ] Advertising
- _____ [ ] Automobiles
- _____ [ ] Depreciation
- _____ [ ] Donations
- _____ [ ] Dues and subscriptions
- _____ [ ] Entertaining
- _____ [ ] Interest
- _____ [ ] Legal
- _____ [ ] Licenses and fees
- _____ [ ] Office insurance
- _____ [ ] Office phone
- _____ [ ] Office rent
- _____ [ ] Office salaries
- _____ [ ] Office utilities
- _____ [ ] Pensions
- _____ [ ] Postage
- _____ [ ] Profit sharing
- _____ [ ] Repairs
- _____ [ ] Small tools
- _____ [ ] Taxes
- _____ [ ] Uncollectible accounts

_____ [ ] **Profit**

Figure 4-10 Overhead checklist

Step 12 - Re-check Your Math and Notes

When you've finished estimating all costs, it's time to begin checking. Back at Step 8 I recommended having someone check your price extensions and totals. Now I'm going to go a step further. Have someone review *all* of your work. Urge them to re-trace every step in the estimate. Trace each number on the estimate form back to its source, either the plans, specs, or your worksheets. When errors are found (and there will be some), the original estimator should initial the correction.

Have someone else go back to the original notes that you took from the specifications in Step 3. That person doesn't have to re-read the entire spec. But they do need to check that everything in the notes has been covered in the estimate.

Now have the checker go through the drawings again. Look for any lines that don't have colored marks on them. If there isn't a mark on a line, be sure there's a reason for the omission. Maybe that's an enlarged view of something that appears elsewhere on the plans. The point is, be sure that nothing has been left out of the estimate.

Compare bids from your selected subs with your own take-off. Be sure they've estimated all the work you expect them to complete.

Only when you're satisfied that everything's accounted for should you go on to the final steps.

Step 13 - Verify Total Cost

Once you've finished the physical check, see if it makes sense. Compare the bottom line for this project to costs for similar jobs you've completed recently. Assuming you made money on those jobs, your bid for this job should be about the same.

I recommend comparing unit costs — such as per square foot or per cubic yard. Suppose you're bidding a 1661-square-foot house and your cost estimate is $95,000. Divide that number by 1661 to calculate the cost per square foot at $57.19. If your recent jobs have run between $55 and $65 per square foot, it's safe to assume your $57.19 cost per square foot is about right.

If the job isn't like work you usually handle, compare your estimate to published figures. The order form at the back of this book lists several cost references that will help you in this. You'll most likely use the *National Construction Estimator* and the *Building Cost Manual.*

Compare square foot costs for most new residential construction. If you're bidding a remodeled kitchen or bathroom, compare square foot costs for similar bathroom or kitchen jobs. But I wouldn't compare new construction costs to remodeling costs. Remodeling costs are higher, sometimes much, much higher per square foot of floor.

In industrial construction, compare cubic foot costs (the cubic feet of space under the roof). For hospital work, compare costs per bed or costs per square foot of patient care area. For hotels, you'd compare cost per room. For condos or apartments, compare costs per living unit.

Cost comparisons will probably build your confidence in the estimate. But what if there's a big difference in costs? Suppose your estimate works out to $45 per square foot and jobs like this have run at least $55 in the past. That should put you on alert. Something is probably wrong.

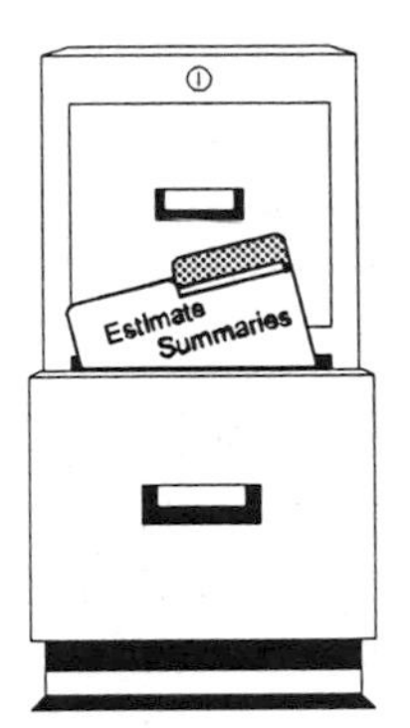

Keep a file of estimate summaries on forms like the one in Figure 4-11. If you prefer, use the more complete estimating forms shown in Figures 4-12 (for industrial and commercial jobs), 4-13 (for plants and equipment) and 4-14 (for residential jobs). Compare your estimate summary for the current bid with the summary of your last similar job. Where are the differences? When you find them, check those areas very carefully.

Maybe this job has a large, unfinished basement. That's much cheaper than a basement with finish flooring and paneling. Maybe your last several jobs included upgrades in glazing and insulation and custom appliances. That may explain the difference.

If you can't find a good reason for the difference, maybe you should start over from scratch. In any case, don't rely on a bid that you suspect is wrong. If it looks wrong, it probably is. Mistakes like that can drastically shorten your career as an estimator.

Step 14 - Add Your Profit

We'll cover both company overhead and profit much more completely later in this book. Chapters 9, 10 and 11 explain how to calculate overhead and profit. I'll also explain why profit has to vary from job to job. For now, we'll just remind you that collecting all the cost figures is never enough. You're never finished until you've added profit and general overhead expenses. Omitting overhead and profit from a bid is like leaving the peanut butter and jelly out of your peanut butter and jelly sandwich. Overhead and profit are your reason for bidding. Yet omitting them from a bid is a common estimating mistake. That may sound funny. But it isn't. It happens all too often, even to contractors who should know better.

It's easy to make mistakes under the stress of a bid deadline, especially if you've had a lot of last minute revisions and corrections. You're less likely to omit markup if you put overhead and profit on your estimating checklist. Then use the checklist on every estimate. Never let a bid leave your office without taking the time to check off every item on the list.

Estimate Summary

Job number: ______________________ **Customer:** ______________________

Estimator: **Date:**

	Cost	Markup	Total
Materials	________	________	________
Plus sales tax		________	________
Equipment	________	________	________
Plus sales tax		________	________
		Material & equipment subtotal	________
Subcontractor	________	________	
Subcontractor	________	________	
Subcontractor	________	________	
		Subcontracts subtotal	________
Labor: Trade	________	________	
Labor: Trade	________	________	
Labor: Trade	________	________	
		Labor subtotal	________
Permits & fees	________	________	
Travel	________	________	
Other expenses	________	________	
		Miscellaneous subtotal	________

	Cost	Markup	Total
Estimate total	________		________
Calculated total markup percentage		________	

Figure 4-11 Estimate summary – short form

Estimate Summary

Owner	Date 19
Owner's address	Telephone
Job address	Lot Blk. Tract

General Conditions	Labor	Other	Total
Supervision:			
Superintendent			
General foreman			
Master mechanic			
Engineer			
Timekeeper			
Assistant timekeepers			
Payroll clerk			
Material checkers			
Watchman			
Waterboy			
Others			
Permits			
Blasting			
Building			
Sidewalk bridge			
Street obstruction			
Sunday work			
Temporary			
Wrecking			
Other			
Bonds			
Completion bond			
Maintenance bond			
Street encroachment bond			
Street repair bond			
Insurance			
Workers' Compensation			
Builder's risk fire insurance			
Completed operations public liability insurance			
Equipment floater insurance			
Public liability insurance			
Truck and automobile insurance			
Licenses			
Local business license			
State contractor's license			
Taxes			
Excise taxes			
Payroll taxes			
Sales taxes			
Field Office			
Owner's			
Job			
Maintenance of office			
Telephone			
Owner's			
Job			
Job office supplies			
Shanties			
Storage			
Tool			
Watchman's			
Transportation of equipment			
Delivery charges			
Travel expenses			
Temporary utilities			
Power			
Light			

General Conditions	Labor	Other	Total
Subtotal forward			
Temporary utilities (cont.)			
Water			
Heat			
Fuel			
Temporary toilets			
Surveys			
Photographs - damage & progress			
Lost time, weather, etc.			
Removing utilities			
Cutting & adjusting for subcontract.			
Repairing damage			
Patching after subcontracting			
Cleanup, general			
Clean windows			
Clean floors			
Cleanup after subcontracting			
Removing debris from job			
Signs			
Pumping			
Protection of construction			
Protection of adjoining land and buildings			
Barricades			
Temporary fences			
Special Conditions			
Plant and equipment. (See checklist)			
Acoustical			
Air conditioning			
Architectural concrete			
Architectural terra cotta			
Bins			
Cabinetwork			
Bookcases			
Cases			
Coat closets			
Counters			
Displays			
Linen closets			
Metal cabinets			
Special cabinets			
Stationery cabinets			
Store fixtures			
Telephone booths			
Wardrobes			
Wood cabinets			
Other			
Cellular-steel floors			
Cofferdams			
Concrete			
Admixtures			
Columns			
Curbs and gutters			
Flatwork			
Floors			
Footing			
Foundations			
Frost protection			

Figure 4-12 Industrial and commercial estimate summary

Special conditions	Labor	Material	Total
Subtotal Forward			
Concrete (cont.)			
Pumpcrete			
Sidewalks			
Slabs			
Vacuum concrete			
Walls			
Walks			
Other			
Concrete curing compounds			
Contingencies			
Conveyors			
Corrugated steel			
Roofing			
Siding			
Culverts			
Concrete box			
Concrete pipe			
Corrugated metal			
Metal arch			
Doors			
Exterior wood			
Panel			
Flush			
Dutch			
French			
Exterior metal			
Revolving			
Kalamein			
Overhead			
Garage			
Screen			
Tin clad			
Industrial			
Other			
Glass			
Plastic			
Interior metal			
Shop			
Office			
Interior wood			
Panel			
Flush			
Sliding			
Frames			
Metal			
Wood			
Electric fixtures			
Electric wiring			
Building wiring			
Service			
Power wiring			
Motors			
Electric signs			
Other			
Elevators			
Excavation			
Clearing and grubbing			
Removing obstructions			
General excavation			
Blasting			
Shoring			
Structural excavation			
Trench excavation			
Backfill			
Rough grading			
Fine grading			

Special conditions	Labor	Material	Total
Subtotal forward			
Rock			
Fences and Railing			
Fire alarm			
Fireproofing			
Flooring			
Asphalt tile and linoleum			
Composition			
Cork			
Flagstone			
Hardwood			
Marble			
Rubber			
Tile			
Slate			
Terrazzo			
Formwork			
Arches			
Beams			
Beam and slab floors			
Bridge piers			
Caps			
Columns			
Fiber tubes			
Flat slabs			
Floor pans			
Footings			
Foundation walls			
Girders			
Metal pans			
Movable forms			
Other			
Foundations			
Wall footings			
Piers			
Spread footings			
Piles, bearing			
Wood			
Steel			
Concrete			
Piles, sheets			
Wood			
Steel			
Concrete			
Caissons			
Frost protection			
Hardboards			
Hardware			
Rough			
Finish			
Heating			
Incinerator			
Insulation			
House			
Cold storage			
Piping			
Rigid			
Flexible			
Foil			
Other			
Lath			
Metal			
Ceilings			
Exterior-walls			
Interior-walls			
Soffits			
Partitions			

Figure 4-12 (cont.) Industrial and commercial estimate summary

Special Conditions	Labor	Material	Total
Subtotal Forward			
Lath (cont.)			
Beads			
Plaster boards			
Walls			
Ceilings			
Arches			
Lift slab			
Loading dock			
Lumber construction			
Heavy			
Beams			
Blocks			
Braces			
Built-up beams			
Caps			
Centering			
Columns			
Falsework			
Girders			
Girts			
Joists			
Lagging			
Laminated members			
Plank and laminated floors			
Planking			
Plates			
Posts			
Sheeting			
Stringers			
Timber purlins			
Trusses			
Trussed beams			
Wales			
Wedges			
Light			
Beams			
Blocking			
Bracing			
Bridging			
Columns			
Cripples			
Dormers			
Fascia			
Furring			
Girders			
Half timber work			
Headers			
Hips			
Jacks			
Joists, floor, roof & ceiling			
Outlooks			
Pier pads			
Plates			
Plywood, flooring, sheathing and roofing			
Posts			
Rafters			
Ribbons			
Ridges			
Roof trusses			
Sheathing-roof, wall			
Sills			
Steps			
Studs			
Subfloor			
Trimmers			

Special Conditions	Labor	Material	Total
Subtotal forward			
Marble			
Manholes			
Masonry			
Ashlar			
Common brick			
Face brick			
Concrete blocks			
Precast concrete panels			
Clay tile			
Dimension stone			
Rubble stone			
Flue			
Metal work			
Art metal work			
Base			
Casings			
Chair rail			
Column guards			
Cornices			
Elevator entrances			
Fire escapes			
Freight doors			
Grillwork			
Information boards			
Linen chutes			
Lintels			
Mail chutes			
Metal doors and frames			
Platforms			
Railings			
Shop front			
Shutter			
Stairs			
Treads			
Trap doors			
Transoms			
Wheel guards			
Wainscoting			
Other			
Millwork and Finish			
Base			
Built-ins			
Casings			
Caulking			
Corner boards			
Cornice			
Doors			
Frames			
Jambs			
Mantels			
Molding			
Paneling			
Sash			
Screens			
Shelving			
Siding			
Sills			
Stairs			
Stops			
Storm doors			
Trim			
Windows			
Wood carving			
Other			
Painting and Decorating			
Aluminum paint			

Figure 4-12 (cont.) Industrial and commercial estimate summary

Special Conditions	Labor	Material	Total
Subtotal Forward			
Painting and Decorating (cont.)			
Doors and windows			
Finishing			
Floors			
Lettering			
Masonry and concrete			
Metal			
Paperhanging			
Plaster			
Roofs			
Shingle stain			
Stucco			
Wood			
Pavements			
Asphalt			
Base			
Block			
Brick			
Concrete			
Wood			
Other			
Pipelines			
Cast iron pipelines			
Concrete pipelines			
Corrugated pipelines			
Steel pipelines			
Trenches			
Vitrified tile pipelines			
Plaster			
Bases			
Coves			
Cement			
Exterior			
Interior			
Interior			
Keene's cement			
Models			
Ornamental			
Perlite			
Special finish			
Plumbing			
Interior			
Exterior			
Prestressed concrete			
Railroad work			
Reinforcing steel			
Bars			
Mesh			
Spirals			
Stirrups			
Retaining walls			
Roofing			
Asbestos			
Asphalt shingles			
Built-up			
Concrete			
Copper			
Corrugated			
Aluminum			
Asbestos			
Steel			
Gravel			
Gypsum-poured and plank			

Special Conditions	Labor	Material	Total
Subtotal Forward			
Roofing (cont.)			
Shingles			
Slate			
Steel			
Tile			
Tin			
Sandblasting			
Service lines			
Sheet metal			
Shutters			
Skylights			
Sound deadening			
Stacks			
Stalls			
Structural steel			
Anchors for structural steel I			
Bases of steel or iron			
Beams, purlins and girts			
Bearing plates and shoes			
Brackets			
Columns of steel, iron or pipe			
Crane rails and stops			
Door frames			
Expansion joints			
Floor plates			
Girders			
Grillage beams			
Hangers of structural steel			
Lintels			
Monorail beams			
Painting steel and remove rust			
Rivets			
Steel stacks			
Welding			
Engineering and shop details			
Inspection			
Freight			
Unloading			
Tanks			
Test holes			
Thresholds			
Tile			
Tilt-up concrete			
Trench shoring			
Vaults and vault doors			
Wallboards			
Waterproofing			
Well drilling			
Windows			
Frames			
Wood			
Steel			
Other			
Sash			
Wood			
Steel			
Aluminum			
Subtotal:			
Contingency			
Profit			
Total:			

Figure 4-12 (cont.) Industrial and commercial estimate summary

Plant and Equipment Checklist			
Air compressors		Pumpcrete equipment	
Asphalt plants		Pumps	
Asphalt tools		Radio units	
Backhoes		Rollers and compactors	
Bulldozers		Grid rollers	
Cables		Power rammers	
Clamshells		Rubber-tired rollers	
Concrete buckets		Sheepsfoot rollers	
Concrete chutes		Steel rollers	
Concrete hoppers		Vibrator compactors	
Concrete mixers		Salamanders	
Conveyors		Saws, power	
Cranes		Scaffolds	
Crawler		Scrapers	
Truck		Shop equipment	
Diving apparatus		Shores	
Distributors		Small tools	
Draglines		Surveyor's instruments	
Drilling equipment		Tractor shovels	
Electrical generators		Tractors	
Elevating graders		Crawler	
Hoist equipment		Rubber tired	
Hooks		Shovels	
Hoses		Trailers	
Jack hammers		Travel-loaders	
Jacks		Trenchers	
Lift-and-carry cranes		Trucks	
Lift trucks		Vibrators	
Light plants		Wagons	
Motor graders		Water trucks	
Night flares		Welding units	
Power shovels		Wellpoints	

Figure 4-13 Plant and equipment checklist

Residential Estimate Summary							
Owner				Contractor			
Property address				Address			
Estimate by				Telephone number			
Legal description				Net Bldg. Fund		Loan Number	

No. Item	Qty	Unit	Material	Labor	Subcontract	Total	Actual cost
1 Excavation							
2 Trenching							
3 Concrete rough							
4 Concrete finish							
5 Asphalt paving							
6 Rough lumber							
7 Finish lumber							
8 Door frames							
9 Windows and glass							
10 Fireplace							
11 Masonry							
12 Roof							
13 Plumbing							
14 Heating							
15 Electric wiring							
16 Electric fixtures							
17 Rough carpentry							
18 Finish carpentry							
19 Lath and plaster							
20 Drywall							
21 Garage door							
22 Doors							
23 Painting							
24 Cabinets							
25 Hardwood floors							
26 Ceramic tile							
27 Formica							
28 Hardware							
29 Linoleum							
30 Asphalt tile							
31 Range and oven							
32 Insulation							
33 Shower door							
34 Temperature facilities							
35 Miscellaneous							
36 Cleanup							
37 Carpet							
38 Gutters and downspouts							
39 Septic tank							
40 Sewer connection							
41 Overhead							
42 Supervision							
43 Water meter							
44 Building permit							
45 Plan and specifications							
46							
47 **Total Cost**							

Figure 4-14 Residential estimate summary

How do you compute the markup? There are a couple of ways to do the math. One way is to multiply the job cost by the markup percentage and then add that result to the job cost. For example, suppose the job cost is $100,000 and you want to add a markup of 25 percent:

✏ $100,000 x 0.25 = $25,000

✏ $100,000 + $25,000 = $125,000

If you don't need to see markup in dollars as a separate step, there's an easier way. Just multiply your cost by 1 plus the markup percentage converted to a decimal. Using the same $100,000 job cost and 25 percent markup:

✏ $100,000 x 1.25 = $125,000

Notice that both of these methods produce the same answer. But does the $125,000 bid price yield a 25 percent margin on the job? Certainly not! The margin ($25,000) is only 20 percent of the selling price ($125,000).

✏ $25,000 ÷ $125,000 = 0.2 or 20%

If you want your margin to be 25 percent of the selling price, use a different procedure. Subtract your margin (expressed as a decimal) from 1.00. Then divide the result into your cost. Using the same figures:

✏ 1.00 - 0.25 = 0.75

✏ $100,000 ÷ 0.75 = $133,333

Check your math to be sure:

✏ $133,333 x 0.25 = $33,333

✏ $33,333 ÷ $133,333 = 0.25, or 25%

Step 15 - Check the Bid Package

This step is essential on all government jobs. If you bid on these projects, you'll have to submit everything exactly the way the contracting agency or officer requires it. Otherwise your bid won't be considered. It will probably be thrown out as "non-responsive."

To avoid disappointments, check the *Instructions to Bidders* one more time before you seal that envelope. Be sure you've completed all the forms, included all the required information, and signed your name every place you're supposed to.

Imagine how you'd feel at bid opening if you discovered the company's bid bond in your pocket instead of in the envelope. Too late, you realize you forgot to put it in the bid envelope. First you hear that your bid is the apparent winner. Then you hear that your bid has been disqualified. Nothing can be more discouraging to an estimator. Any disqualification on a technicality is going to be a disappointment. A little carelessness can destroy the product of hours of hard work.

Technicalities aren't observed so precisely in private work. Most owners will accept a phone quote with a written confirmation to follow. If something was omitted, no problem. Just send it over when you can. If yours is the best price, you'll get the job. That's usually all there is to it.

The Detailed Estimate Advantage

You may not always prepare detailed estimates on every job. Even highly experienced estimators will cut some corners occasionally, especially if the markup is high enough to cover mistakes. But every professional estimator has to know how to prepare accurate, detailed estimates, especially in a tight market when the difference between high and low bids is only a few percent.

If your estimates require accuracy only to the nearest 10 percent, you can probably do most estimates in an hour using the square foot method. But it takes more than an hour to compile an estimate that's accurate to within a few percent.

A 20% margin for error is unusual

When all contractors have as much work as they can handle, estimates may include a 20 percent margin for error. Most contractors will stay in business even without accurate estimates. But times like that don't last. Soon enough, careless estimators with marginal skills get squeezed out. Don't be one of these. Good estimating practice can help you survive in the lean times and thrive in the good times.

I recommend that you follow the fifteen steps I've outlined here when you prepare any significant estimate. You can't afford to estimate by guessing. Accuracy takes time. But it also produces the best results. That's especially important when there's plenty of work to be done and good markups are available.

In the next two chapters I'll explain in detail the heart of construction cost estimating, the take-off.

Chapter 5
Estimating Repair & Remodeling Work

In the first chapter I promised to make this book useful to all estimators, no matter what you're estimating. Now I have to deviate a little — this chapter is strictly for remodeling estimators. If you don't estimate remodeling or repair work, and you don't want to learn how to estimate this type of work, move right along to the next chapter. There's not much here for you. We cover estimating procedures for new construction in detail in Chapter 6.

But if you already estimate repair and remodeling work, or want to get into this field, this chapter deserves close attention. Here's why. Estimating remodeling work is fundamentally different. It isn't necessarily harder than estimating new construction. It's just different. True, the materials are about the same as in new construction. And the same construction trades are required. But what gets estimated is very different.

Every remodeling job includes demolition and removal. Estimated quantities are usually much smaller. Most important, you have to make things fit and match, and you must work around existing construction. These conditions require that you improvise much more than on new work.

Also, there are more surprises. Problems with the building inspector are routine. The potential for disputes with the owner lurks in every wall. The job will require more labor per operation than similar new construction.

And there usually won't be complete plans for the job. Often, you'll be the one to draw the plans and write the specs. All this increases your risk and makes it harder to predict costs.

You can estimate remodeling jobs accurately

In fact, I've heard remodelers claim that there's no way to estimate remodeling costs accurately. I don't agree. There *is* a way, and I'm going to show you how to do it. My way requires that you:

- Become thoroughly familiar with the job
- Find out about the existing conditions
- Pay attention to detail

If you use the system described here, you'll produce more accurate, more consistent estimates. I'll explain how to be sure your bid covers the work to be done. I'll also explain how to avoid losses on unknown and unforeseen conditions.

Limiting the Scope of Your Bid

So what's a remodeling estimator supposed to do? Fortunately, it's not as hard to compile good repair and remodeling estimates as you might expect. As in all kinds of estimating, you try to anticipate the most likely problems. Unless you're clairvoyant, you can't anticipate every possible repair and remodeling problem. So the focus of every estimate has to be on ways to protect yourself from costs that you can't forecast. One way to do this is called *limiting the scope of your bid.*

Avoid disputes

Your bid has to identify exactly what's *included in* and what's *excluded from* your estimate. Anticipate what might cause disputes and specifically exclude those items from your estimate. And never describe in the estimate what the owner will end up with. Instead, describe the materials you're going to furnish and install. That's an important distinction. Let me explain with an example.

There are two ways to estimate a roofing job. In new construction it's okay to bid a "300 lb. composition shingle roof." You know what the shingles cost and how long installation will take. The owner will end up with a composition shingle roof. That's easy. But let's compare that to a re-roofing job.

Suppose your bid says "Re-roof with 300 lb. composition shingles." When you start the job, you discover that the roof deck is rotted out over the porch and won't even hold a nail. You have to replace the rotted deck before you do any other work. Worse yet, the roof is so weak it's not even safe to walk on in some places. Several decayed rafters should be replaced. And when you pull off the flashing, some of it gets damaged, so you can't use it on the new roof. The cost of this extra work may be more than the cost of the shingles. Who pays? The owner should, of course. You know that. The owner gets the benefit of the extra work. You bid only the shingles. But try telling that to the owner.

The owner's going to insist that he accepted your bid for a new roof. And that's exactly what he wants, a new roof — and at the price you bid. "Not a penny more," he says.

Of course, you could drive nails into that rotted deck and hope the rafters hold up a little longer. And maybe you can find some way to re-use the flashing. But the first windstorm would scatter shingles all up and down the street. And the rusted, twisted flashing is going to look terrible on that new deck. You take too much pride in your work (and your reputation) to do shoddy work just because the owner won't pay for essential extras. So you "bite the bullet" and do what has to be done — at a loss — and chalk it up to experience, and bad estimating.

If that's what you've been doing, this chapter should keep you from ever doing it again. It's not especially hard to avoid a loss like that. It just takes a slightly different approach to estimating and bidding.

Protecting Yourself

As I said, your estimates and bids have to be very clear about what's included and what's excluded. The owner wants a new roof. But sometimes you can't be sure what a new roof requires until work begins. So you don't bid a "new roof." Instead, you "furnish and install" materials. For example, your proposal should say "Furnish and install 1600 square feet of 300 lb. composition shingles." If that makes a new roof, fine. If a new roof requires replacing roof deck, several new rafters and some flashing, that's okay too. But you bid only the shingles. Everything else is at extra cost.

To protect yourself, follow these three rules of repair and remodeling estimates.

Three rules for clean estimates

Rule 1. Describe exactly what you plan to remove or demolish.

- How much?
- What type?
- What's included?
- What's excluded?
- What are you going to do with the debris?
- What about salvage value?

Offer your best efforts to limit damage to adjacent surfaces. Explain that you can't guarantee that the petunias under the soffit won't get trampled and the dog won't get loose. Better that the owner transplant the petunias and tie up the dog before you start work.

Rule 2. Describe very clearly the *materials* you plan to furnish and install.

- How much?
- What type?
- What brand?
- What accessories will be included?

It should be easy to identify these materials. They're what you plan to buy for the job. Don't even mention the finished product, such as a "new roof."

Rule 3. Exclude by name the most likely extras. You know what those are. On a re-roof job, they're:

- Sheathing
- Rafters
- Flashing
- Vents
- Waterproofing
- Repair to skylights

Also exclude anything the building inspector may require that isn't listed under Rule 2 above. You can't be responsible if the inspector decides that it's time to upgrade the electric service panel.

If you think these exclusions will make it hard to sell the job, there's an easy way to make them "inclusions." Just list prices for all the likely extras. For example, your bid could specifically exclude replacement of roof sheathing but show a cost per panel or per square foot *if* you have to replace sheathing.

You're Not an Insurance Company

That's the key to better repair and remodeling estimates. You're an estimator. You're estimating the cost of furnishing and installing construction materials. You're not an insurance company. Insurance companies are in business to make good on losses. If something bad happens and you've got insurance, great. You're covered. That's why you buy insurance.

Contractors aren't licensed to write insurance. They're licensed to construct buildings. If the inspector demands a change, or if there's a vent stack in that partition you planned to move, or if there's any kind of surprise — no problem! You're happy to bid on that work too. But it's not your loss. If it's not on the list of materials you agreed to furnish and install (your estimate), it's *not part of the contract.*

A Sample Remodeling Estimate

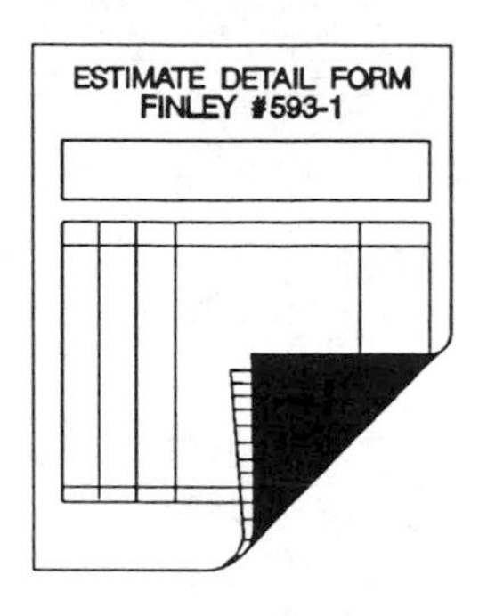

Learning to estimate remodeling work is like learning to ride a bike. I could explain bicycle riding for two weeks and you wouldn't learn as much as in the first three minutes of trying. The best way to learn estimating is to try it. So let's take a little practice spin.

I'm going to walk you through a sample estimate step-by-step. This is a small kitchen remodeling project, but the procedures would be the same for a bathroom job, an attic or basement conversion, a room addition or even repair after a fire loss.

Kitchens Plus

As the name suggests, Kitchens Plus Remodeling specializes in kitchen remodeling. The owner is Rick Miller. He's a *licensed* general contractor and a good finish carpenter with over 15 years' experience in residential construction and remodeling. Rick uses a crew of three on most jobs: himself, Joe, a journeyman carpenter, and Mike, a helper. Like most remodelers, the Kitchens Plus crew handles work that fits under many different trade classifications.

For example, the crew of three usually does all the framing, drywall, tile setting, and finish carpentry. They install windows, doors and cabinets. They also do minor electrical work, plumbing and HVAC work. (Relocating or adding to existing plumbing, electrical or HVAC systems normally doesn't require a licensed subcontractor.) Rick usually hires a specialty contractor for major new plumbing, electrical or HVAC work, even on a remodeling job.

Rick is the chief salesman, does all the estimating, prepares bids and supervises the work. A bookkeeping service does the accounting, makes out the payroll and prepares tax returns. Kitchens Plus is run from Rick's home. His wife answers the company phone. She pays company bills and sets up sales calls. Rick does most of his business with four local building material dealers and has good working relationships with all of them.

Kitchens Plus has been in business a little over three years. Rick is proud of his reputation for doing excellent work, on time and for the contract price. He tries to do the job right the first time and has very few callbacks. At least half his new jobs come through referrals from past customers.

Sales for last year were a little over $300,000. Price for a typical job is from $5,000 to $7,000. Kitchens Plus averages about one job a week and keeps the crew of three busy nearly all year. Overhead (mostly Rick's salary as manager) runs about 20 percent of the bid price and the company usually makes a reasonable profit on each job.

To make 20 percent on the bid price, Rick uses a 25 percent markup. Here's how the arithmetic works:

Suppose Rick's total estimated costs, including all material, labor, equipment and subcontracts, is $10,000. He adds a 25 percent markup, making the bid price $12,500.

✏ $10,000 x 0.25 = $2,500

✏ $10,000 + $2,500 = $12,500

So Rick's markup on that $12,500 job is $2,500. That's 20 percent of the selling price.

✏ $12,500 x 0.20 = $2,500

Sales have increased each year since Rick started doing business under the Kitchens Plus name. This year sales are up more than 15 percent. Rick's five-year goal for Kitchens Plus is to reach $400,000 in annual sales. He's now three years into that plan and reaching the goal seems likely. To get there, Rick will need to set up a second full-time crew. If Kitchens Plus is to remain profitable and competitive, Rick can't afford to hire a supervisor for that second crew. He feels he can supervise two jobs at once. But he knows running two jobs will reduce the time and attention he can give to each.

The Finley Job

Ed and Bernice Finley got Rick's business card from a Kitchens Plus customer. The Finleys bought an older home and want to remodel the kitchen before their daughter gets married in three months. Rick has an appointment to discuss the project.

Because most Kitchens Plus jobs are small (under $10,000), most owners don't have plans drawn. This was the case when Rick arrived for his sales call, and he prefers it that way. He can develop his own drawings and specifications and control the scope of the job right from the start. Rick prefers to recommend the cabinets and fixtures offered by his regular suppliers. He also likes to deal with homeowners directly, rather than through a designer or architect. Best of all, there's usually no competition. If Rick draws a plan the owners like and submits a proposal they can afford, he usually gets the work without bidding against anyone.

When Rick entered the Finleys' kitchen, he could see that the existing cabinets had been repainted several times and were in poor condition. The sheet vinyl flooring was shabby and the seams were lifting. The sink and faucet were obsolete and the countertop was chipped and scorched in several places. Rick listened to Ed and Bernice explain what they had in mind and asked enough questions to get a good mental picture of their needs.

The Finleys wanted to replace the old wall and base cabinets with more modern units. They wanted a new countertop and a new sink in the counter. Bernice said she would love to have oak plank flooring if it would fit into their budget.

Rick showed them pictures of cabinets, countertops and sink fixtures offered by the dealers he prefers. The Finleys made tentative selections of each. Ed Finley said he could get a dishwasher, refrigerator and a 30-inch range and range hood at cost through his brother-in-law. So Rick didn't have to worry about appliances. Bernice said she would repaint the kitchen after the cabinets are installed. Rick's bid will be very specific. His contract will say that the Finleys are responsible for the appliances and painting, and that those items aren't included in his bid.

Rick suggested a projected garden window over the sink and a luminous drop ceiling with fluorescent fixtures above. The Finleys liked the idea but wanted to wait on that until after the wedding. "That's for phase two," said Ed. Property values are increasing in the Finleys' neighborhood. Ed feels money spent on their home is money well invested. He plans to remodel their two bathrooms next year.

Don't base bids on potential jobs

Rick makes a mental note of that information, but it doesn't influence his bid. He's learned that basing his bids on "potential" jobs doesn't work. He bids each job on its own, at a fair price for the work he does. He does good work, and gets a lot of repeat business.

Discussing Price

Ask the right questions

As Rick measured the kitchen and sketched the existing layout, Ed asked about the cost. *Don't answer that question!* Rick said he would prepare a plan and a detailed proposal that would fit within the Finleys' budget. Ed's question gave Rick the opening to ask a key question, "How much do you want to spend?" Ed didn't answer with an amount, but said they had some money set aside and could borrow more if they had to.

Rick replied that he worked with a major bank in the area, and would be happy to help them meet the loan officer at the local branch. This bank has a good home improvement loan program and this is the type of loan they like to make. Ed said they would prefer not to borrow, if possible. They had $6,000 set aside. If the cost wasn't much over that, they wouldn't need a loan. Rick said he could work within that budget.

The Finleys told Rick they had no use for the old cabinets, so he should figure the cost of hauling them away.

Before he left, Rick reviewed his notes with the Finleys to be sure the scope of work was right. He told them the Kitchens Plus bid would include:

Removal and disposal of:
- Existing kitchen cabinets
- Existing countertops and sink
- Existing vinyl flooring

Furnish and install:
- Kitchen cabinets and countertops
- Stainless steel kitchen sink and trim
- Either new vinyl flooring or oak plank flooring
- Patching of gypsum wallboard damaged by removal work
- Cleanup of debris caused by construction

Submit your bids quickly

The Finleys agreed that Rick's notes were correct. Then Rick asked Ed what other company was going to bid on the job. Ed said he wanted to get several quotes but hadn't contacted anyone yet. He mentioned the names of two other remodeling companies. Rick considered both of them to be legitimate competitors. He thanked the Finleys for their time and promised he'd be back with a written proposal within three working days. He explained that the proposal would include a floor plan and elevation views so they could see how the kitchen would look when the job was done. He would also quote an exact price, set a date when work would start, and estimate the completion time.

Job Site Investigation

Before he left the Finley home, Rick thought about how he'd handle the materials for this job. There were only two ways to get to the kitchen. One was through the front door and down a hall with two 90-degree turns. The other way was in a side gate, around the garage, across the back yard, and through a door on the back porch. He chose the back porch entrance, even though it required extra handling of materials. He noted this on his clipboard sketch and left after about 90 minutes at the Finleys' home.

The Estimating Process

Start with a sketch

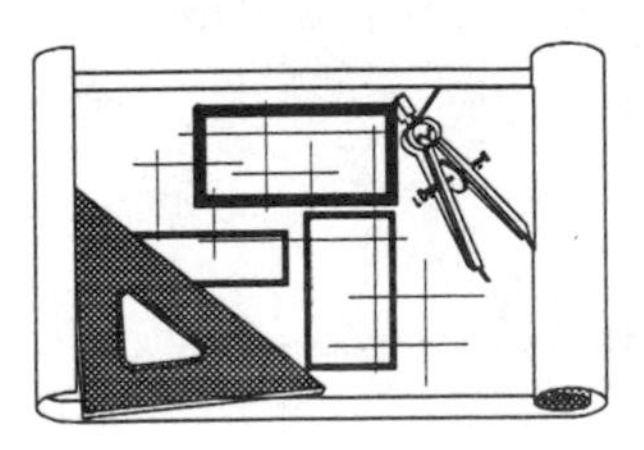

Back at his desk, Rick sat down to prepare a drawing of the proposed kitchen. He used the rough sketch he made during his visit with the Finleys to redraw the room size, and locations for the windows, doors, and appliances. He worked to fit the best combination of cabinets into the available space. Figure 5-1 shows the plan Rick prepared. Notice that Figure 5-1 also includes a bill of materials that lists each of the items shown on the drawing. This bill of materials defines the scope of work for the Finley job.

Proposed Kitchen Design
Prepared for Ed and Bernice Finley
Prepared by: Kitchens Plus Remodeling
Estimate #593-1 Date: April 9, 1993
Refer to Job Specifications

Bill of materials

	Qty.	Description
1	1	15" base cabinet
2	1	15" 4-drawer base cabinet
3	1	24" sink base cabinet
4	1	21" base cabinet
5	1	42" blind corner base cabinet
6	1	24" base cabinet
7	1	36" base cabinet
8	2	18" 4-drawer base cabinet
9	2	21" wall cabinets
10	1	24" wall cabinet
11	1	15" x 30" wall cabinet
12	2	36" wall cabinet
13	1	15" x 36" wall cabinet
	20 LF	Seamless countertop w/4" backsplash
	84 SF	Oak plank or sheet vinyl flooring
	1	Single stainless steel sink and trim

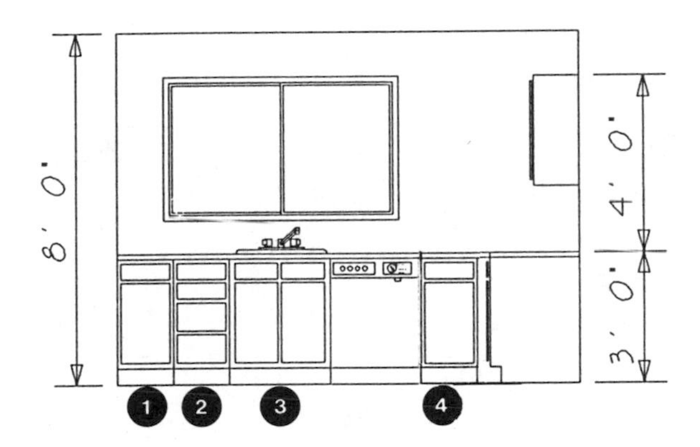

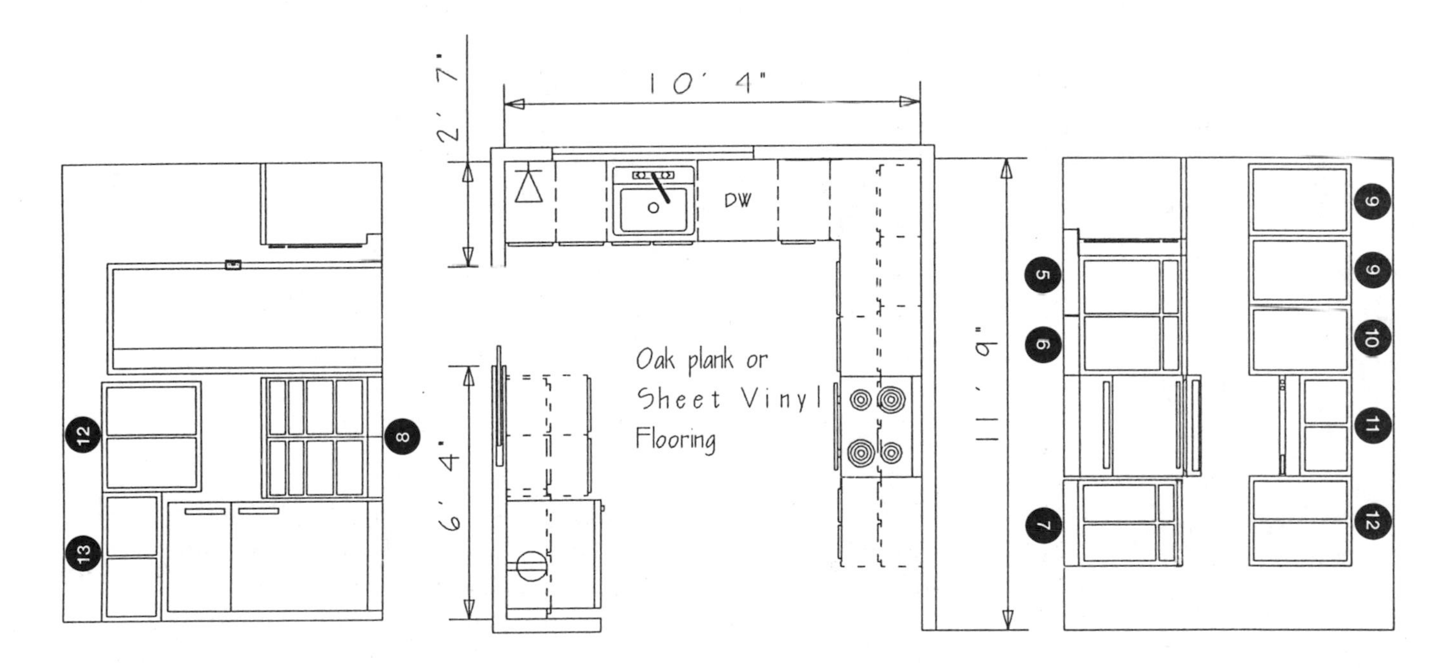

Figure 5-1 Kitchen plan and elevation views

Estimate Detail Form

Company: Kitchens Plus Remodeling | **Estimate #** 593-1 | **Estimator:** Rick Miller | **Date:** 4/7/93

Job: E&B Finley kitchen remodel | **Estimate due:** 4/9/93 | **Checked by:** Connie Miller | **Date:** 4/8/93

Address: 123 Main Street, Anytown

Job Description: Remove & replace cabinets, sink, etc.

CSI Division/Account: 2 / Site preparation & removal work

Notes: See layout sketch, by Rick

Use a two-man crew: Joe Vargas, carpenter at $23 per hour & Mike Goss, helper at $19 per hour = $21 per manhour average

Item or Description	Qty	Unit	Material Unit $	Material Ext $	Manhours MH/Unit	Manhours MH Ext	Labor MH $	Labor Ext $	Equipment Unit $	Equipment Ext $	Subcontract Unit $	Subcontract Ext $	Total Cost
REMOVE EXISTING CABINETS, ETC.													
No salvage value, no re-use													
Wood construction, installed on													
gypboard/wood stud walls													
Countertop, Marlite	20	LF	--	--	.050	1.00	21.00	21.00	--	--	--	--	21.00
Base cabinets, 24" deep	8	Ea	--	--	.250	2.00	21.00	42.00	--	--	--	--	42.00
Wall cabinets, 12" deep	7	Ea	--	--	.357	2.50	21.00	52.50	--	--	--	--	52.50
Disconnect & remove sink	1	Ea	--	--	.500	0.50	21.00	10.50	--	--	--	--	10.50
Cut drywall, install backing boards													
for new cabinets - 2x6 Allow for	30	LF	0.42	12.60	120	3.60	21.00	75.60	--	--	--	--	88.20
Patch drywall, 1/2" Allow for	100	SF	0.50	50.00	.020	2.00	21.00	42.00	--	--	--	--	92.00
Remove sheet vinyl flooring	100	SF	--	--	.015	1.50	21.00	31.50	--	--	--	--	31.50
Clean up area caused by our work	120	SF	0.05	6.00	Allow	1.00	21.00	21.00	--	--	--	--	27.00
Load items removed on our truck,													
haul to dump, pay dump fee	1	LS	Allow	20.00	Allow	2.00	21.00	42.00	Allow	50.00	--	--	112.00
Total Direct Costs this sheet, Removal work			Total Material $	88.60	Total Manhours	16.10	Total Labor $	338.10	Total Equip $	50.00	Total Subcont $	--	Total $ 476.70

Wage rates used include base pay, payroll taxes, insurance & fringe benefits. Overhead & profit NOT INCLUDED

Base pay: carpenter = $16.00 & helper = $13.50 | Use 16 manhours for scheduling | Estimate Detail Form 1 of 5

Figure 5-2 Estimate detail, demolition

Furnish a drawing with your proposal

Rick will include a copy of the plan with his proposal to the Finleys. This drawing will help the Finleys visualize what they're getting and should help close the sale. Rick will use the same drawing when he estimates labor and material costs.

Once he finished the drawing, Rick began his estimate. He started by assigning a name (E & B Finley kitchen remodel) and number to this job. See Figure 5-2. Since this job would be the first scheduled for completion

Job Estimate Log

Month: April 1993 Page 1 of __

Job Number	Bid Due Date & Estimator	Job Name & Address	Description of Work	Bid Amount	Comments
593-1	4/9/93 Rick	E&B Finley	R&R Cabinets		
		123 Main St.	flooring, sink		

Figure 5-3 A page from Rick's estimate log

Number your estimates

in May of 1993, Rick assigned the estimate number 593-1. He'll use this job name and number on all the paperwork related to the Finley job, such as material purchase orders, work schedules, and manhour records.

Rick wrote the job address, a short job description, his name and the estimate date at the top of the form. Since removal work falls under Construction Specifications Institute (CSI) Masterformat Division 2, Site Preparation, he enters, "2 Site Preparation/Removal" after CSI Division/Account on the fifth line of the form.

Log your estimates

Next, Rick wrote information about this estimate in the bid log book and noted the bid due date. The Kitchens Plus bid log is a three-ring binder where Rick records key facts about each bid. Figure 5-3 shows part of the bid log page that lists the Finley job. When the estimate is complete, Rick will write the bid price in this log. Figure 5-4 is a blank bid log that you can copy and use. Notice that Rick uses preprinted forms for his estimates. Figure 5-5 is the blank Estimate Detail Form Rick uses. You're welcome to reproduce this form also.

Now, Rick is ready to begin the estimate. He'll use the sketch and notes he made at the Finleys', and estimate the job in the same sequence as his crew will do the work. He'll work through the job step-by-step in his mind, and write down each step on the estimate sheet as he goes. It's less likely that he'll forget something if he follows this sequence.

Job Estimate Log

Month: __________ Page __ of __

Job Number	Bid Due Date & Estimator	Job Name & Address	Description of Work	Bid Amount	Comments

Figure 5-4 Blank estimate log sheet

Estimate Detail Form

Company: ______________________________
Job: ______________________________
Address: ______________________________
Job Description: ______________________
CSI Division/Account: __________________

Estimate # __________
Estimate due: ________

Estimator: _____________ Date: ________
Checked by: ___________ Date: ________

Notes:

Item or Description	Qty	Unit	Material		Manhours		Labor		Equipment		Subcontract		Total Cost
			Unit $	Ext $	MH/Unit	MH Ext	MH $	Ext $	Unit $	Ext $	Unit $	Ext $	

Total Direct Costs this sheet	Total Material $	Total Manhours	Total Labor $	Total Equip $	Total Subcont $	Total $

Estimate Detail Form __ of__

Figure 5-5 Blank estimate detail

Out With the Old

The *Item or Description* column

The first tasks will be to remove the old countertop, cabinets, sink and vinyl flooring. Back in Figure 5-2, page 1 of the Estimate Detail Form for this job, you can see that Rick began with the demolition and patching work. He listed each task or item in the *Item or Description* column.

He also used that column to note that the estimate assumes no salvage or re-use of materials and that cabinets are hung on wood-frame walls covered with gypsum wallboard. Notes like these make this estimate more valuable months or years later. They identify the work in detail. If actual costs are more or less than estimated, Rick will want to know why. The notes may help explain it. He prefers not to make the same mistakes twice.

Rick listed nine items in the *Item or Description* column, starting with the countertop and ending with loading debris on the company truck. Notice that he left a blank line between most work items. There are two reasons for leaving those blank lines. First, they make the estimate easier to read. If you fill every line, the estimate turns into a sea of numbers and letters. An estimate like that is hard to read and understand, especially for someone who isn't a construction estimator. Second, this estimate will be reviewed many times before the Finley job is finished. Blank lines leave room for changes, additions and notes.

Changes to the estimate would be easier to make if Rick used a computer for estimating. Although Rick uses a computer to write letters and keep track of his bank account, he doesn't use it for estimating. Rick feels writing estimates by hand is easiest. He may be right. Most experienced estimators still do most of their estimating with pencil and paper. Of course, it's much easier to make changes, add columns, summarize totals, and print extra copies with a computer. Decide for yourself which is best after you read Chapter 7 and experiment with *National Estimator.*

The *Qty* & *Unit* columns

After Rick listed the nine demolition and patching tasks, he began to enter quantities and units of measure for each of them. The countertop is 20 feet long. He entered 20 in the *Qty* (Quantity) column and LF in the *Unit* column. On the row opposite "Countertop," he wrote "Marlite." As you see, he wrote quantities and units of measure for each task on page 1 (Figure 5-2) of the estimate.

***Material cost estimates* -** Next, Rick began estimating material costs. Removal work usually doesn't call for many materials. But he figures he'll need about 30 feet of 2 x 6s as backing to hang the new cabinets, and he'll have to patch about 100 square feet of drywall after he removes the existing cabinets.

Rick's material cost estimate for the 2 x 6 backing is 42 cents per linear foot (LF) for the material.

Lumber is often priced per thousand *board feet*, abbreviated *MBF*. A board foot is the same as 1 square foot of lumber, 1 inch thick. To calculate the board feet per linear foot of lumber, multiply the nominal width times the nominal depth, *in inches*, and divide the answer by 12. Thus, a 2 x 6 board is equal to 1 board foot per linear foot.

✏ 2 x 6 = 12

✏ 12 ÷12 = 1

In this case Rick's price was $420 per MBF.

You can calculate the cost per linear foot of any piece of dimensioned lumber the same way. For example, let's see how much per linear foot a 2 x 4 at $420 per MBF would be.

✏ 2 x 4 = 8

✏ 8 ÷12 = 0.667 (two-thirds of a board foot)

Now multiply the $420 per MBF by the 0.667. The total cost for 1000 linear feet is $280.14. Divide $280.14 by 1000, and you have 28 cents per linear foot for 2 x 4s at $420 per MBF.

Rick likes to work with costs of lumber based on a price per linear foot. Your lumber supplier can quote you a lumber price per linear foot based on specific sizes and lengths. When it comes time to ship the lumber materials to the job, Rick will order it in lengths that will minimize waste.

The *Unit $* column

The wallboard, tape, joint compound and Spackle he'll need for patching is 50 cents per square foot (SF) of wall. Notice that in the *Unit $* column under *Material*, Rick wrote 42 cents per linear foot for the 2 x 6s and 50 cents per square foot to patch 100 square feet of drywall. Then he estimated 5 cents per square foot for cleanup. That's the cost of solvent and sandpaper to clean adhesive residue off the 120 square feet of floor (the total floor area) after he pulls up the existing flooring and removes the old cabinets.

He'll base the cost of materials and installation for the new floor on the net area to be covered after the new cabinets are installed. The above costs are all the material costs for the removal part of the job.

The *Ext $* column

List the extended materials cost in the *Ext $* column under *Material*. Rick multiplied the cost of the 2 x 6s (42 cents) by the quantity (30 linear feet) to get the extended cost of $12.60. He did the same for drywall patching and cleanup. Material cost for debris disposal is estimated as a lump sum (LS in the *Qty* column). The cost is $20, which is Rick's estimate of the charge at the landfill dump.

Total material cost will be $88.60 for the removal work. Notice 88.60 at the bottom of the *Ext $* column.

The *MH/Unit* column

Manhour estimates **-** Next, Rick filled in manhour estimates in the column headed *MH/Unit*. He estimated only 0.05 manhours per linear foot to remove the countertop. He knows from records of completed jobs that it would take more time if the top had to be salvaged (used again). But in this case, salvage isn't required. His crew can cut the top in two, if necessary, to make the job quicker and easier.

Notice that Rick estimates removal of base and wall cabinets as separate items. That's because he knows from experience that it usually takes his crew longer to remove a wall cabinet (0.357 hours, or about 21 minutes) than it does to remove a base cabinet (0.25 hours, or 15 minutes). He can't use the same manhour rate for both.

Some estimators prefer to figure cabinet demolition by the linear foot of cabinet face (or back) rather than by the number of cabinets. That's because it might be inaccurate on some jobs to estimate demolition by number of cabinets. For example, the manhour estimate (per cabinet) for a kitchen with many small cabinets might be too generous. But an estimate based on footage might also be too high if the cabinets are unusually large. Here's a case where you have to rely on your own experience and judgment.

Notice that Rick has estimated the manhours required for every removal task in this job. He's included allowances in his estimate for disconnecting the sink, pulling up the sheet vinyl flooring, cleaning up the work area, loading debris on the company truck and hauling to the dump. How can Rick be sure of these estimates? The simple answer is that he can't. But he knows these tasks will take *some* time. If he omits any of them from his estimate, his estimated time and cost for that item is zero. That's going to be wrong in every case. Better to allow what seems to be a reasonable time rather than no time at all.

Where does Rick get these manhour figures? His best source of reliable manhour estimates is his own history of completed jobs. Chapter 8 describes a quick and simple way to compile manhour productivity records from your construction and remodeling projects. Remember, there should be at least two profits in every job. One goes into your bank account. The other is what you learn about estimating that type of work. Every professional estimator should keep good cost records on completed jobs. Rick does. And he uses them regularly when he estimates work like the Finley job. If you need some guidance on how to build up a good set of cost records, order a copy of *Cost Records for Construction Estimating*. There's an order form bound into the back of this book.

Assigning manhours and costs **-** Where do you get good manhour estimates if you don't have years of experience estimating remodeling costs? That's a problem for every beginning estimator. When you've completed dozens of jobs, nearly every estimate is similar in some ways to an estimate you've done before. But until then, what are you supposed to do?

Fortunately, many published sources are available. In fact, *National Estimator* (which is included on the CD-ROM) includes over 2,500 pages of manhour estimates for all types of building construction. You'll see these manhour estimates when we get to Chapter 7. Of course, no published reference manual will have estimates for your crews on your jobs using your supervisors. Only you have that information. But if nothing else is available, published data may be your only option.

No matter what estimating data is available, you still need to rely on your good common sense. Assign what you feel are reasonable manhour values to any estimate. If it makes sense, it's probably right — or at least close enough to avoid a major loss. The more estimates you compile, the more confidence you'll develop.

Experienced estimators will tell you, "If you don't know the cost of something, better guess to the high side." As you get better at analyzing required manhours, you'll be able to reduce the allowance you need to cover unknowns. The more guesses you can eliminate, the more realistic and competitive your estimates and bids should be.

The *MH Ext* column

Extending labor costs **-** To continue with his estimate, Rick multiplied the manhours per unit by the number of units to find the extended manhours. That's the figure in the column headed *MH Ext* for each of the nine cost items in Figure 5-2. Notice that the sum of all extended labor estimates is 16.10 manhours. That figure is in the *Total Manhours* box at the bottom of the *Manhours* columns. Rick wrote "Use 16 manhours for scheduling" at the bottom of the form, because he prefers to work with whole numbers when he estimates crew time. He didn't change the total labor of $338.10 to $336.00 (16 manhours times the $21 per manhour) because he wants to check his arithmetic across each row as well as down each column.

The *Labor MH $* column

The next column Rick completed is headed *Labor MH $*. This is the average hourly labor cost for the two-man crew that will do the removal work. Notice the hourly wage rates listed on the sixth line of this estimate. The hourly labor cost is $23 for Joe (the carpenter) and $19 per hour for Mike (the helper). These hourly costs include the base wage ($16.00 for Joe and $13.50 for Mike), fringe benefits (like vacation pay), taxes (like F.I.C.A.) and insurance (like Workers' Comp).

The average hourly cost per manhour for this two-man crew is $21 ($23 plus $19, or $42 divided by 2). Notice that $21 is listed for each of the nine removal tasks in the *Labor MH $* (labor cost per manhour) column. This is not the *crew* cost per hour. It's the average cost per *manhour*. That's appropriate because all the time estimates on the form are per manhour, not per crew hour.

The *Labor Ext $* column

The Labor Ext $ column **-** Next, Rick multiplied the hourly labor cost ($21) by the manhours required for each removal task. He entered this figure in the column headed *Labor Ext $*. Once Rick entered all the labor costs, he wrote their total ($338.10) at the bottom of the column.

As a quick check of his addition, Rick multiplied the manhour total (16.10 hours) by the average hourly rate ($21) to get the same figure, $338.10. That's called using a *crossfoot* to check column addition. That's a good idea for every page where the hourly labor cost is the same for every item. Make it a habit to check column totals. Remember that the most common mistake on estimates is in the addition. Use the crossfoot total to spot errors before they become expensive losses.

The *Equipment* and *Subcontract* columns

Equipment and subcontract costs - The next columns to the right are used for equipment and subcontract cost estimates. For this job, the only equipment cost is for use of the company truck to haul debris to the dump. Rick allowed $50 for this. In the next chapter, and also in Chapter 9 I'll discuss in more detail how to charge for equipment. For now, just remember to charge each job an hourly rate for equipment, whether you own it or rent it. If you rent the equipment, write the rental cost in the *Equipment* column, and don't forget to include any charges for getting the equipment to your job and back to the yard again.

If you or one of your workers has to go to the rental yard to pick up or return equipment, be sure to include the time (manhours) and cost (labor) to cover this. Estimate the cost of fuel for gas- or diesel-powered equipment, and don't overlook per-mile charges for vehicles.

There were no subcontract costs for this part of the job. But when you have a job where you'll use subcontractors, briefly describe the work in the *Description* column. Write the quantity (usually 1) and unit (usually LS for lump sum) in their respective columns. Then write your chosen subcontractor's quote in the *Subcontract* column.

Totals and Schedules

The *Total Cost* column

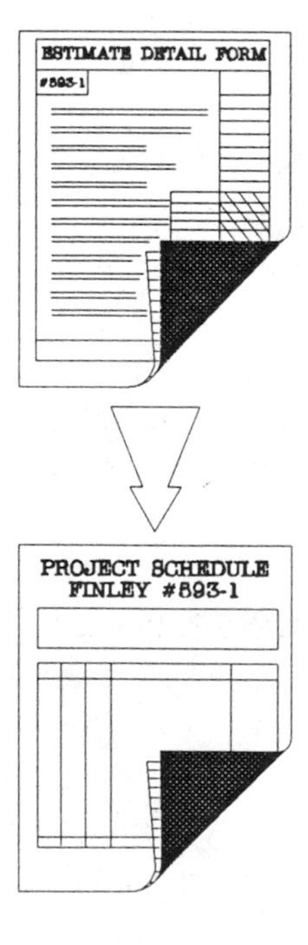

Rick's next step was to add all the cost figures for each line and write the sum in the *Total Cost* column. Then he added up that column. Next, he checked that number against the sum of the column totals. They should be the same. Rick's total estimated cost for removal is $476.70. Now he can start thinking about scheduling the work.

Notice that Rick expects to use a crew of two for removal. His estimate is that the job will take 16 manhours. A manhour is one person working for one hour. A crew of two working an 8-hour day would put in 16 manhours (2 men times 8 manhours each per day is 16 manhours). So for this crew, it will take one day to complete the removal. Rick had two reasons for doing that calculation.

First, he wanted to be sure that his estimate for removal was realistic. He decided it was. A crew of two should be able to take out the cabinets, countertop and flooring on a small job like that in about one day. It made sense to him, so it was probably right.

Rick has learned to be practical when he estimates manhours. Most crews will try to be at the end of some part of the job at quitting time. Sometimes that means dragging out the work so a task doesn't get done until 4 p.m. Other times the crews will have to work quickly to finish by 4. Either way, Rick's goal is to make manhour estimates match actual production. Of course, he tries to have plenty of work available for his crew every day. But that isn't always possible. If a job requires a little less than a full day's work for a crew, he increases the manhour estimate to show a full day's work anyway.

A half-day minimum charge is realistic

Rick knows from experience that it isn't practical to schedule any job for less than a half-day minimum for a crew. He knows that a half-day minimum charge is realistic because the crew takes just as long to set up and break down for a two-hour job as they do for an eight-hour job. The only time he'll schedule less than half a day is for a callback to fix something or respond to a customer's complaint. And then he tries to combine callbacks so he can schedule a full day for the crew to do several callbacks at once. That way he avoids interfering with his other job schedules.

Rick expects his crew to spend the first day on the Finley job doing removal. They'll be ready to install cabinets on the morning of the second day. Rick makes sure the cabinets are delivered before the scheduled installation date.

Estimating the Cabinets

Figure 5-6 (Furnish and Install New Cabinets) shows Rick's cabinet installation estimate. It's labeled *Estimate Detail Form 3 of 5* in the lower right-hand corner. Rick followed exactly the same estimating procedure on this page as he followed on page 1, the removal estimate (Figure 5-2). He completed each column, starting with the column at the left. At the bottom of each cost and manhours column he entered the column total. Whenever possible, he checked the crossfoot to be sure he added correctly.

A few points on Figure 5-6 are worth special attention:

Key estimates to drawings

First, notice the numbers down the left margin. These refer to the cabinets as they're labeled on the sketch in Figure 5-1. On this Estimate Detail Form, Rick listed the cabinets the way they're arranged in the manufacturer's catalog and price list. This isn't the same order in which he assigned his numbers on the drawing. There, he started in the upper left-hand corner on the plan with the Number 1 and worked his way

Estimate Detail Form

Company: Kitchens Plus Remodeling — **Estimate #** 593-1 — **Estimator:** Rick Miller — **Date:** 4/7/93

Job: E&B Finley kitchen remodel — **Estimate due:** 4/9/93 — **Checked by:** Connie Miller — **Date:** 4/8/93

Address: 123 Main Street, Anytown

Job Description: Remove & replace cabinets, sink, etc.

CSI Division/Account: 12 / Furnishings - new cabinets

Notes: See layout sketch by Rick. Cabinet # refers to Bill of Materials

Use a two-man crew: Joe Vargas, carpenter at $23 per hour & Mike Goss, helper at $ 19 per hour = $21 per manhour average

Item or Description	Qty	Unit	Material Unit $	Material Ext $	Manhours MH/Unit	Manhours MH Ext	Labor MH $	Labor Ext $	Equipment Unit $	Equipment Ext $	Subcontract Unit $	Subcontract Ext $	Total Cost
Furnish & install new cabinets													
Mill-made modular kitchen cabinets													
(includes hinges & pull hardware)													
per specifications													
Base units 34½" high x 24" deep													
#2 = four drawer, 15" wide	1	Ea	150	150.00	.638	0.638	21.00	13.40					163.40
#1 = One door, one drawer, 15" wide	1	Ea	142	142.00	.638	0.638	21.00	13.40					155.40
#8 = Four drawer, 18" wide	2	Ea	160	320.00	.766	1.532	21.00	32.17					352.17
#4 = One door, one drawer, 21" wide	1	Ea	132	132.00	.461	0.461	21.00	9.68					141.68
#6 = One door, one drawer, 24" wide	1	Ea	170	170.00	.911	0.911	21.00	19.13					189.13
#7 = Two door, two drawer, 36" wide	1	Ea	256	256.00	1.350	1.350	21.00	28.35					284.35
#3 = Special sink base, 24" wide,													
Two door, two drawer fronts	1	Ea	142	142.00	.740	0.740	21.00	15.54					157.54
Allow for added door	1	Ea	30	30.00	.250	0.250	21.00	5.25					35.25
#5 = One door, blind corner, 42" wide	1	Ea	180	180.00	1.500	1.500	21.00	31.50					211.50
Wall units - 15" high x 12" deep													
#11 = Two door, 30" wide	1	Ea	111	111.00	.461	0.461	21.00	9.68					120.68
#13 = Two door, 36" wide	1	Ea	122	122.00	.638	0.638	21.00	13.40					135.40
Wall units - 30" high x 12" deep													
#9 = One door, 21" wide	2	Ea	119	238.00	.638	1.276	21.00	26.80					264.80
#10 = One door, 24" wide	1	Ea	128	128.00	.766	0.766	21.00	16.09					144.09
#12 = Two door, 36" wide	2	Ea	174	348.00	1.030	2.060	21.00	43.26					391.26
***Subtotal all manhours above						**13.221**							
Allow for extra handling	25	%				3.305	21.00	69.41					69.41
Total Direct Costs this sheet, New cabinets			Total Material $	2469.00	Total Manhours	16.526	Total Labor $	347.06	Total Equip $	--	Total Subcont $	--	Total $ 2816.06

Wage rates used include base pay, payroll taxes, insurance & fringe benefits. Overhead & profit NOT INCLUDED

Base pay: carpenter = $16.00 and helper = $13.50 — Use 16 manhours for scheduling — Estimate Detail Form 3 of 5

Figure 5-6 Estimate detail, cabinets

clockwise around the plan. He listed the base cabinets first, then the wall units. These reference numbers make it easy for anyone to understand the estimate — Rick, the Finleys, and the cabinet installation crew.

Reference numbers make omissions easy to spot

Reference numbers also make it easy to spot an omission. To make sure he had accounted for all the cabinets on the layout sketch, Rick compared the Estimate Detail Form to the bill of materials on the sketch. Beginning with #1, he marked each cabinet on the estimate form with a pencil check mark, then marked it on the drawing also. When he checked them off on the layout sketch, he also verified that the quantities matched in both places. Anyone checking the estimate can see immediately which cabinet goes where on the plan — and will probably spot a mistake if there is one. This method works as well for large estimates, which may span several pages, as it does for this one.

Mark drawings in color

I suggest you use a similar method to mark items on your estimates. Use colored pencils to mark each item on the drawing or bill of materials as you list it on your estimate. Make it as easy as possible to understand and check your estimates.

Second, you may have noticed that we skipped page 2 of Rick's estimate. Rick estimated the job in the order he expected to do the work: removal, cabinets, countertop, sink and then flooring. So he finished pages 1, 3, 4, and 5 of the estimate before he began page 2, the flooring estimate. *Estimate Detail Form 2 of 5* appears later in this chapter as Figure 5-9. When Rick had finished all five pages, he reassembled them in Masterformat order and numbered the pages:

Arrange your estimate in Masterformat order

- Division 2, Removal, page 1 of 5
- Division 9, Flooring, page 2 of 5
- Division 12, Cabinets, page 3 of 5
- Division 12, Countertop, page 4 of 5
- Division 15, Sink, page 5 of 5

This way, his complete estimate is arranged in Masterformat order. On a larger job there may be several pages of Estimate Detail Forms for each Division. And each major Division will likely be broken down into subcategories. When you have more than one Estimate Detail Sheet for a given CSI category (or subcategory), include the CSI account number at the bottom of each detail sheet. Follow that with "Page x of x" for that account. Then, summarize each account on a recap sheet. The recap should have a line for each page, with the totals from that page.

Total each account recap sheet, and carry those totals forward to the Estimate Summary Sheet, which we'll discuss in a moment.

Third, note the last line on page 3 of Rick's estimate (Figure 5-6). It shows an additional 25 percent added to the labor time for handling cabinets through the gate, around the garage and in the back door. The notes he made when he visited the job site reminded him to allow extra handling time. In this case, increasing the installation time by 25 percent brought manhours for hanging the cabinets to about 16 hours. That's convenient. It's exactly one day's work for Rick's two-man crew. When day two on the Finley job is over, the cabinets should be in place and ready for countertops.

When Rick finished page 3, he considered for a moment whether the manhour estimate for handling and hanging cabinets was realistic. His judgment was that the estimate was about right. A crew of two should be able to lay out, handle, install and fit-up 16 cabinets in one day, even with the extra time required to carry cabinets through the gate and back yard. Remember that Rick already included time to install the backing (2 x 6 boards) in his estimate for removal.

Also, notice the line below item #3, the Special Sink Base, that says "Allow for added door." That's because a standard 24" sink base cabinet from this manufacturer normally comes with one door. There's an extra charge to supply this cabinet with two doors.

Countertop and Sink Estimates

Pages 4 and 5 (Figures 5-7 and 5-8) show estimates for the countertops and the single-compartment stainless steel sink. A cabinet shop gave Rick a bid of $605 for the tops. He estimates that installation, including time for fit-up router work, will take about 6 manhours. See Rick's note about installing the flooring and sink on the same day as the countertops. He realized that 6 manhours is only 3 hours for the two-man crew. He wants to schedule a full day's work for that crew, if possible, since they're the same ones who will install and connect the sink.

Rick's estimate of the time to install the sink and trim is 2 manhours. (See Figure 5-8). He also included a note on this sheet about installing the countertops and flooring on the same day as the sink. Consistency pays off, and a little duplication doesn't hurt. These notes make understanding and explaining Rick's estimates much easier.

The Flooring Option

Bernice Finley wants oak plank flooring if the cost fits her budget. Rick will take several samples of oak plank with him when goes back to the Finleys' to present the Kitchens Plus bid. But if he can't sell the oak plank floor, he'll fall back to his bid for sheet vinyl. Take a look at Rick's note about "Option #2" on Figure 5-9. He indicates that sheet vinyl flooring instead of oak plank will reduce the bid by $500. He also carries this information forward to the Estimate Summary sheet (Figure 5-10).

This is a small flooring job at less than 90 square feet. No matter which floor covering the Finleys select, installation won't take more than 8 manhours. Note that Rick plans to use the same two-man crew for this work. We know from the note on Figure 5-9 that Rick's estimate includes floor preparation that wasn't part of demolition on the first day of the job. It also includes adhesive. His estimate of 8 manhours for the flooring work makes sense. So, it will take 16 manhours to install the countertops, sink and flooring. That's an eight-hour day for Rick's crew of two.

Estimate Detail Form

Company: Kitchens Plus Remodeling | **Estimate #** 593-1 | **Estimator:** Rick Miller | **Date:** 4/7/93

Job: E&B Finley kitchen remodel | **Estimate due:** 4/9/93 | **Checked by:** Connie Miller | **Date:** 4/8/93

Address: 123 Main Street, Anytown

Job Description: Remove & replace cabinets, sink, etc.

Notes: See layout sketch, by Rick

CSI Division/Account: 12 / Furnishings - Countertops

Use a two-man crew: Joe Vargas, carpenter at $23 per hour & Mike Goss, helper at $19 per hour = $21 per manhour average

Item or Description	Qty	Unit	Material		Manhours		Labor		Equipment		Subcontract		Total Cost
			Unit $	Ext $	MH/Unit	MH Ext	MH $	Ext $	Unit $	Ext $	Unit $	Ext $	
Furnish & install new countertops													
One-piece custom countertops													
25" deep with 4" back splash,													
shop laminated plastic on													
composition backing board.													
Solid color with square edge.													
Total 19'-10" (19.83'), use													
20 LF on layout sketch.													
"Sink top" type: L-shaped													
10'-4" long leg	10.33	LF	30	310.00	.290	3.00	21.00	63.00					373.00
Cabinet top, two 36" long	6	LF	30	180.00	.333	2.00	21.00	42.00					222.00
one 42" long	3.5	LF	30	105.00	.285	1.00	21.00	21.00					126.00
Add for sink cut-out by													
manufacturer	1	Job	10	10.00	--	--	--	--					10.00
Fit-up router work included above	1	Job	--	--	--	--	--	--					--
Note: Install flooring and sink on same day as this work is performed See Estimate Detail Form Sheet 2 of 5 and 5 of 5													
Total Direct Costs this sheet, New countertops			Total Material $	605.00	Total Manhours	6.00	Total Labor $	126.00	Total Equip $	--	Total Subcon $	--	Total $ 731.00

Wage rates used include base pay, payroll taxes, insurance & fringe benefits. Overhead & profit NOT INCLUDED

Base pay: carpenter = $16.00 and helper = $13.50

Estimate Detail Form 4 of 5

Figure 5-7 Estimate detail, countertops

Putting It All Together

Work systematically on your estimates, as Rick did. Identify each work item, list quantities, estimate installation time, materials, and equipment. Use a separate detail form for each part of the job. Then use an Estimate Summary like the one in Figure 5-10 to gather and total costs from the bottom line of the estimate detail forms. Figure 5-11 is a blank Estimate Summary for you to use.

Estimate Detail Form

Company: Kitchen Plus Remodeling
Estimate # 593-1
Estimator: Rick Miller
Date: 4/7/93
Job: E&B Finley kitchen remodel
Estimate due: 4/9/93
Checked by: Connie Miller
Date: 4/8/93
Address: 123 Main Street, Anytown
Job Description: Remove & replace cabinets, sink, etc.
Notes: See layout sketch, by Rick
CSI Division/Account: 15 / Plumbing
Use a two-man crew: Joe Vargas, carpenter at $23 per hour & Mike Goss, helper at $19 per hour = $21 per manhour average

Item or Description	Qty	Unit	Material Unit $	Material Ext $	Manhours MH/Unit	Manhours MH Ext	Labor MH $	Labor Ext $	Equipment Unit $	Equipment Ext $	Subcontract Unit $	Subcontract Ext $	Total Cost
Furnish & install new kitchen sink													
Countertop opening for sink will													
be pre-cut													
Bill of materials: (See labor													
following materials)													
Single bowl, stainless steel,													
self-rimming sink, 24" x 21" deep	1	Ea	73	73.00									73.00
Deck mounted faucet, polished													
chrome w / spray attachment	1	Ea	105	105.00									105.00
Drain fitting, polished chrome with													
tail piece and removable strainer	1	Ea	16	16.00									16.00
Hot & cold flex supply connectors													
with angle stop valves	1	Set	21	21.00									21.00
Trap assembly	1	Ea	30	30.00									30.00
Labor to install all items above													
and connect piping to existing													
plumbing at wall at rear of sink	1	Job	--	--	2.00	2.00	21.00	42.00					42.00
Note: Install countertops and flooring on same day as this work is performed													
See Estimate Detail Forms Sheets 2 of 5 and 4 of 5													
Total Direct Costs this sheet, plumbing kitchen sink			Total Material $ 245.00		Total Manhours 2		Total Labor $ 42.00		Total Equip $ --		Total Subcon $ --		Total $ 287.00

Wage rates used include base pay, payroll taxes, insurance & fringe benefits. Overhead & profit NOT INCLUDED

Base pay: carpenter = $16.00 and helper = $13.50

Estimate Detail Form 5 of 5

Figure 5-8 Estimate detail, sink

Index your estimate

Use a numbering system such as the Masterformat account numbers for each part of your job. Put your estimate detail page number on each line of the estimate summary. That makes it easy to check any figure on the estimate summary.

After you've copied column totals from all the detail sheets to your summary, add the columns and write their totals on the "Subtotal, direct job costs" line. The number in the *Total Cost* column should agree with the

Estimate Detail Form

Company: Kitchens Plus Remodeling | **Estimate #** 593-1 | **Estimator:** Rick Miller | **Date:** 4/7/93

Job: E&B Finley kitchen remodel | **Estimate due:** 4/9/93 | **Checked by:** Connie Miller | **Date:** 4/8/93

Address: 123 Main Street, Anytown

Job Description: Remove & replace cabinets, sink, etc.

Notes: See layout sketch by Rick

CSI Division/Account: 9 / Finishes/Flooring

Use a two-man crew: Joe Vargas, carpenter at $23 per hour & Mike Goss, helper at $19 per hour = $21 per manhour average

Item or Description	Qty	Unit	Material Unit $	Material Ext $	Manhours MH/Unit	Manhours MH Ext	Labor MH $	Labor Ext $	Equipment Unit $	Equipment Ext $	Subcontract Unit $	Subcontract Ext $	Total Cost
Furnish & install wood plank flooring													
Existing sheet vinyl was stripped as part of removal work													
Pre-finished oak strip flooring,													
tongue-and-groove $\frac{25}{32}$" thick, random plank $2\frac{1}{4}$" to $3\frac{1}{4}$" wide, installed over existing concrete slab (after cabinets are installed).													
Quantity shown includes 8% allowance for waste													
Prep floor, clean and scrape	84	SF	--	--	Allow	2.00	21.00	42.00					42.00
Plank flooring	84	SF	7.26	610.00	Allow	6.00	21.00	126.00					736.00
Adhesive (req'd per manufacturer)	84	SF	0.89	75.00									75.00
Option #2: Sheet vinyl flooring in lieu of oak plank: no wax, Armstrong, Timespan, .80" gauge, @$12 SY Allow 10 SY = $120 vs $610; **DEDUCT** = $332 plus markup = allow $500													

Note: Install countertops and kitchen sink on same day as this work is performed.

See Estimate Detail Form sheets 4 of 5 and 5 of 5

See Option #2 on Estimate Summary for vinyl flooring in lieu of oak plank. Labor will remain unchanged.

Total Direct Costs this sheet, Oak plank flooring	Total Material $	Total Manhours	Total Labor $	Total Equip $	Total Subcon $	Total $
	685.00	8	168.00	--	--	853.00

Wage rates used include base pay, payroll taxes, insurance & fringe benefits. Overhead & profit NOT INCLUDED

Base pay: carpenter - $16.00 and helper = $13.50

Estimate Detail Form 2 of 5

Figure 5-9 Estimate detail, flooring

sum of the column totals. Now, go back and add the final total in the *Total Cost* column from each of the detail sheets. That number should also agree with the figure in the *Total Cost* column on the "Subtotal" line. If it doesn't, you probably copied something wrong to the Estimate Summary.

The Finley estimate is short. A larger job might include dozens of pages of estimate detail sheets. But no matter how long the estimate, the estimating procedure is very nearly the same.

Estimate Summary Form

Company: Kitchens Plus Remodeling **Estimator:** Rick Miller **Date:** 4/7/93

Job: Ed and Bernice Finley **Checked by:** Connie Miller **Date:** 4/8/93

Address & phone number: 123 Main Street, Anytown 12346 Phone: 555-1234 **Estimate #** 593-1

Job Description: Kitchen remodel, cabinets, countertops, sink and flooring **Estimate due:** 4/9/93

CSI Division / Account: as listed below. Job scheduled completion is within 5 working days after start of work

Estimate Detail Page & Account	Item Description	Material Cost	Labor Manhours	Labor Cost	Equip Cost	Sub-contract Cost	Total Cost
Page 1 Acct. 2	Remove and dispose of cabinets, countertops, sink, floor covering;						
	patch walls, prep walls and floor	88.60	16	338.10	50.00	--	476.70
Page 2 Acct. 9	Furnish and install oak plank flooring	685.00	8	168.00	--	--	853.00
Page 3 Acct. 12	Furnish and install new kitchen cabinets	2469.00	16	347.06	--	--	2816.06
Page 4 Acct. 12	Furnish and install custom countertops	605.00	6	126.00	--	--	731.00
Page 5 Acct. 15	Furnish and install kitchen sink, faucet & trim	245.00	2	42.00	--	--	287.00
All items above	Subtotal, direct job costs	4092.60	48	1021.16	50.00	--	5163.76
Markup	Supervision, overhead & profit	--	--	--	--	--	1290.94
	Total	Total amount of this bid, Option #1					6454.70
		Total amount of this bid, Option #2					5954.70

Figure 5-10 Estimate summary, company copy

Overhead and Profit

Like most construction estimators, Rick adds markup at the end of his estimate. Notice the line "Supervision, overhead and profit." He added $1,290.94, which is 25 percent of the total of the material, labor and equipment costs. This is Rick's usual markup. It covers his expenses for supervision and overhead, and leaves a reasonable profit after he pays all the bills for this job. Maybe most important, he knows that Kitchens Plus is both profitable and competitive at this markup.

For a larger job that involves more risk or uncertainty, he may add a contingency allowance of 5 percent or more. But he anticipates no problems with this job and feels confident that his estimate covers all important costs.

Estimate Summary Form

Company: ______________________________ Estimator: ______________ Date: ______

Job: ______________________________ Checked by: ______________ Date: ______

Address & phone number: ______________________________ Estimate # ______________

Job Description: ______________________________ Estimate due: ______________

CSI Division / Account: ______________________________

Estimate Detail Page & Account	Item Description	Material Cost	Labor		Equip Cost	Sub-contract Cost	Total Cost
			Manhours	Cost			
	Subtotal, direct job costs						
	Supervision, overhead & profit						
	Total						

Figure 5-11 Blank estimate summary

Non-Productive Labor Costs

Many estimators include non-productive labor (such as supervision) as a separate item on their estimates. Rick elected to include non-productive labor as part of his overhead expense on the Finley job. There's nothing wrong with that. The important thing is that you account for non-productive labor somewhere on every estimate. It's a major cost on nearly every job.

Estimate non-productive labor

Non-productive labor on site includes all the minor tasks besides installing construction materials:

- Buying and delivering materials
- Planning the work
- Setting up for work each day
- Cleaning up at the end of each day

If you keep close track of the time you spend on tasks like these, you can list them on your estimate as a separate item.

To simplify your estimates, I recommend that you not define productive labor too narrowly. For example, I consider all the following to be *productive labor*:

- Unloading tools and materials
- Measuring
- Taking a coffee break
- Discussing what to do next
- Keeping the job cleaned up as the work progresses.

An estimator knows his workers do more than just pound nails and saw lumber during their "productive" hours. You're safe to assume that there won't be more than about 50 minutes of real work in any productive manhour. Don't expect much more. This doesn't mean that you increase your manhour estimate to allow for the normal non-productive time that's related to construction work. What it means is that your manhour figures should already have non-productive time built into them.

You can include off-site productive labor (usually called *shop labor*) in your estimate as material cost. For example, if Rick builds the cabinets in his garage, that's shop labor. He'd put the labor and material cost to build those cabinets in the materials column of the Finley estimate. Use the labor column only for on-site labor.

Checking the Estimate

When Rick finished the estimate, he laid it aside on his desk and asked his wife to check the arithmetic. She did so, and then wrote her name and the date at the top of each page of the estimate.

Double-check your arithmetic

Always have someone other than the estimator check the arithmetic. If you made a math error once, you're likely to make the same error again. Get a second opinion and you'll eliminate nearly all the simple math errors.

When Rick's wife isn't available to check estimates, Rick has his lead carpenter, Joe, check them. Joe likes to help with the estimates occasionally. And Rick likes to get Joe involved in a new job while it's still in the planning stage — before he bids the work. There's another benefit, too. Employees feel more like a member of the team when they review manhour estimates before the bid is submitted. For Rick, the estimate is a forecast. But for Joe, it's a quota. The estimate shows how much work he's supposed to do in a day. People naturally have more respect for standards they help set, or at least can review before they're imposed. Getting Joe to approve a bid in advance makes it more likely that Joe will finish the work on schedule.

Preparing the Bid

Once Rick finished the estimate, he got ready to submit the bid. When both Rick and his customers sign the bid, it becomes a valid construction contract. How much of this detailed cost estimate will Rick present to the Finleys when he goes to get the contract signed?

Rick likes to keep it simple, but not too simple. Some remodelers enclose the entire cost estimate with their bid. Rick thinks that gives the wrong impression and opens up a troublesome opportunity. He doesn't want the Finleys to break down the quoted costs to find parts of the job they want completed or omitted. That's not the purpose of an estimate. Rick proposes to do the entire project for a set price. He doesn't want to quibble about individual costs. Rick feels that giving the Finleys a copy of his entire estimate would invite quibbling, comparison shopping and nitpicking.

Some remodelers include almost no detail in their bids. The client gets only a blank contract that lists the work to be done and the price. There's no cost breakdown at all. Take it or leave it. Rick feels that limits his flexibility. He may need to do a little horse trading, cutting a corner here and giving a discount there, to sell the job. That's easier if both he and the client have some cost detail to work with. Besides, Rick likes to explain how he estimated the job. That eliminates potential problems. It's important that the Finleys know exactly what's included and what's excluded before they sign the agreement.

Estimate Summary Form

Company: Kitchens Plus Remodeling **Estimator:** Rick Miller **Date:** 4/7/93

Job: Ed and Bernice Finley **Checked by:** Connie Miller **Date:** 4/8/93

Address & phone number: 123 Main Street, Anytown 12346 Phone: 555-1234 **Estimate #** 593-1

Job Description: Kitchen remodel, cabinets, countertops, sink and flooring **Estimate due:** 4/9/93

CSI Division / Account: as listed below. Job scheduled completion is within 5 working days after start of work

Estimate Detail Page & Account	Item Description	Material Cost	Labor Cost	Equip Cost	Subcontract Cost	Total Cost
Page 1 Acct. 2	Remove and dispose of cabinets, countertops, sink, floor covering;					
	patch walls, prep walls and floor	88.60	338.10	50.00	--	476.70
Page 2 Acct. 9	Furnish and install oak plank flooring	685.00	168.00	--	--	853.00
Page 3 Acct. 12	Furnish and install new kitchen cabinets	2469.00	347.06	--	--	2816.06
Page 4 Acct. 12	Furnish and install custom countertops	605.00	126.00	--	--	731.00
Page 5 Acct. 15	Furnish and install kitchen sink, faucet & trim	245.00	42.00	--	--	287.00
	Subtotal, direct job costs	4092.60	1021.16	50.00	--	5163.76

Work Excluded:
- Painting (Customer will do)
- Electrical (Existing)
- Appliances (Customer will furnish and install)
- Walls, doors, windows (existing)

Estimate Summary Page 1 of 1

Supervision, overhead & profit	1290.94
Total amount of this bid, Option #1	6454.70
Total amount of this bid, Option #2	5954.70

Figure 5-12 Estimate summary, customer copy

Rick prefers to include only the Estimate Summary with his proposal. It shows major parts of the job, the Kitchens Plus markup and the total job cost. It has just enough detail, not too little and not too much. It describes materials to be furnished and installed, with quantities. And it includes notes about what's excluded from the job. That helps prevent later disputes about the job scope.

If you show estimated manhours on the Estimate Summary you give your customer, you might create the problem of having to justify them. Better to leave that column off the customer's copy, as in Figure 5-12.

When Rick returns to the Finley home, he'll spend most of his time discussing the plans he drew, and the Estimate Summary. This gives Rick a chance to display both his professionalism and his sincerity. Rick's clients

Kitchens Plus Remodeling
1234 Contractor Road, Anytown, USA 12345
Phone: KIT-CHEN (548-2436)

April 6, 1993

Mr. and Mrs. Ed Finley
123 Main Street
Anytown, USA 12346

Dear Ed and Bernice:

I am pleased to submit this proposal for remodeling the kitchen in your home at 123 Main Street. Work can begin within 10 working days after you accept this proposal. Your new kitchen can be ready for use with 5 working days after start of construction. We work very hard to minimize your inconvenience during the construction period.

Kitchen Plus proposes to furnish and install materials and do the work described below:

- Removal and disposal of:
  - existing kitchen cabinets and countertops
  - existing sheet vinyl flooring
- Furnish and install:
  - kitchen cabinets and countertops
  - stainless steel kitchen sink and trim
  - prefinished oak plank flooring or vinyl flooring
  - patching of gypsum wallboard damaged by removal work
  - cleanup of debris caused by our work

Plans, a bill of materials and job specifications are submitted with this proposal and describe the work we propose to do.

We will complete this work for the sum of $ 6454.70

We require payment of 5% on approval of this proposal and 50% when work begins. The balance will be due on completion of our work. This proposal is subject to approval within 10 days from today. Your signatures below indicate your acceptance of this proposal and your authorization to begin work.

Thank you for the opportunity to submit this proposal. I am looking foward to doing this work for you. Please feel free to let me know if you have any questions or need additional information.

Accepted on: ____________________

______________________ ______________________

Ed Finley Bernice Finley

____

Richard Miller, *Kitchens Plus*

Figure 5-13 Proposal letter

want to have confidence in his ability. They want to believe that what he proposes will match their expectations. Rick develops that confidence by showing that he's done a first-rate job on the estimate and bid.

Prepare a written proposal

Figure 5-13 is the written proposal Rick prepared for the Finleys. It refers to the bill of materials, drawings, and job specifications. That way, everything written on the bill of materials, plans and job specs becomes part of the signed contract.

Rick's contract identifies the job location, when he can start work, how much it will cost, and how long it will take. The contract limits the time for acceptance to 10 days and includes a payment schedule. For a small job like this, Rick likes to collect 5 percent when his customer signs the contract, and at least one-half when work begins. The balance is due on completion.

I recommend you use a Job Completion and Acceptance Form like the one in Figure 5-14. Fill out this form, listing in very general terms the things you will furnish and install. Go over this form with the customer when you sign the contract. Then, when you finish the job, present this checklist for the customer to sign. At that point, the customer can note anything that isn't complete. Once the customer signs the form, give them a copy and file the original in the Job File. Include a reference to this form on the final bill.

The contract includes a place for both Rick and the clients to sign. Rick keeps the original and gives a copy to the clients.

Rick will leave the Finleys a sales brochure describing the cabinets. He has also written a description of the items he intends to furnish and install. See the job specifications in Figure 5-15.

Closing the Sale

When Rick arrived at his appointment with the Finleys a couple of days later, he had in his briefcase:

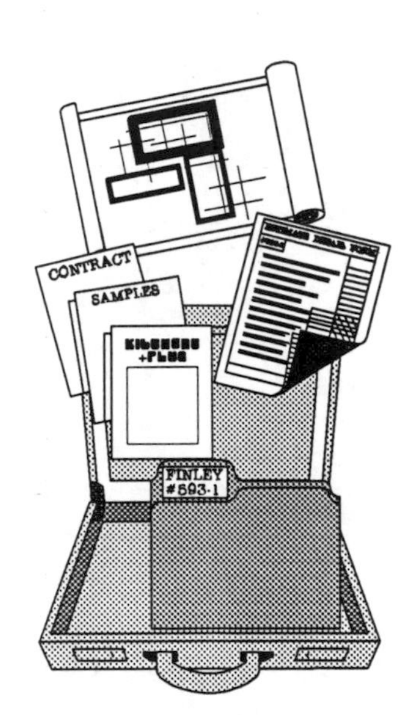

- His plan and elevation views (Figure 5-1)
- The bill of materials (Figure 5-1)
- The Estimate Summary (customer's copy) (Figure 5-12)
- The contract (Figure 5-13)
- A copy of the Job Completion and Acceptance Form (Figure 5-14)
- Specifications for the job (Figure 5-15)
- Oak plank flooring samples
- Vinyl flooring samples
- A brochure describing the cabinets
- A brochure describing his company
- A list of Kitchens Plus references

Rick described the job before he mentioned the price. The Finleys seemed satisfied with his proposal and were obviously relieved when Rick quoted the cost. The price was no surprise. It was very close to the amount they had budgeted. Bernice preferred the oak flooring and knew it would be very attractive with the oak cabinets.

The Finleys weren't alarmed by the price. They didn't attempt to whittle the cost down, and seemed willing to accept the bid as submitted. If they had mentioned other bids, Rick would have urged the Finleys to compare the scope of his proposal with others they had received. "Did those bids include patching the wallboard and preparing the subfloor before

Certificate of Substantial Completion and Acceptance

Project: ______________________ Date of submittal: ______________

Location: ______________________

Contractor: ______________________ Address: ______________________

Owner: ______________________ Address: ______________________

The Contractor considers the work on the Project to be completed in that it is suitable for the Owner to use for its intended purpose. The date of this certificate shall be the date of commencement of all warranties and shall be considered as approval for application for final payment as set forth in the contract. This certificate does not relieve the Contractor from any obligation of the Contract.

Work yet to be completed:

Any other deficiencies or requests for service should be made in writing to the address above. Please be as specific as possible about the defect. Any claim concerning an appliance or fixture under warranty should be made directly to the manufacturer.

I (we) have inspected the work performed and find the job completed to our satisfaction with the exception of the items listed above as yet to be completed.

Date of completion: Owner's signature:

Figure 5-14 Job completion and acceptance form

JOB SPECIFICATIONS

Job Number 593-1
Customer: Ed and Bernice Finley
Prepared by: **Kitchens Plus Remodeling**

Division 2 - Removal work

This work shall include removal and disposing of existing kitchen cabinets, countertops, kitchen sink and kitchen sheet vinyl flooring. All items removed shall become scrap materials with no salvage value and shall be removed from the job site.

Division 9 - Finishes Customer shall select either Option 1 or Option 2

Option 1 Oak Plank Flooring

Prefinished oak strip flooring shall be tongue-and-groove, 25/32" thick random plank 2-1/4" to 3-1/4" wide, number 1 to C and better grade. Installation shall be in accord with the manufacturer's recommended procedure. Flooring shall be installed over an existing concrete slab, properly prepared to receive same using an adhesive as required by the flooring manufacturer.

Option 2 Seamless Vinyl Flooring

Sheet vinyl flooring shall be no-wax Timespan made by Armstrong Flooring Company, .080" gauge. Installation shall be in accord with the manufacturer's recommended procedure. Flooring shall be installed over an existing concrete slab, properly prepared to receive same using an adhesive as required by the flooring manufacturer.

Division 12 - Furnishings

Cabinets

Kitchen cabinets shall be Matador cabinets made by Ariva Cabinet Company described as follows: solid hardwood face frames, hardwood door frames, hardwood drawer fronts, and hardwood veneer on raised door panels (front and back). Joint construction shall be glued mortise, dowel and dado. Cabinets shall have full backs, 1/8" vinyl laminated plywood with vinyl laminated interiors. Cabinet shelves shall be adjustable. Drawer bodies shall be vinyl laminated construction. Cabinet hinges shall be self-closing type. Doors and drawers shall include pull hardware.

Countertops

Countertops shall be one-piece custom-made solid color laminated plastic on composition base with square edges. Backsplash shall be 4" high. Countertops shall be installed the full width and depth on all base cabinets.

Division 15 - Plumbing

Kitchen sink shall be stainless steel, single compartment, self-rimming type, 24" x 21". Sink trim shall include a polished chrome faucet with hand-held sprayer attachment, a removable strainer type drain fitting, hot and cold water valved flexible supply connectors and a drain trap assembly. Drain trap and water supply piping shall be connected to existing piping at the existing wall behind the sink.

Figure 5-15 Job specifications

the finish flooring went down?" Rick would insist that Kitchens Plus bids only a quality job done the right way and with quality materials. And he'd ask the Finleys to consider, "Did the other bidders quote the same?"

Occasionally Rick has to negotiate the price to get a job. But he rarely cuts the price without also cutting something out of the job. Rick likes to think of Kitchens Plus as the value leader in his community. They aren't cutthroat lowball artists. They can compete on price, but aren't willing to sacrifice quality or customer service to do it. Kitchens Plus sells at the Kitchens Plus price, what the job's worth — not the lowest price the client can badger Rick into.

If Rick had sensed that the price was a sticking point, he would have asked if it seemed too high. If the bid Rick submitted had been clearly beyond what the prospect could afford, Rick would have offered to revise the proposal to meet their budget. That would require more estimating time, and he'd have to set up another sales call later. Fortunately, that wasn't the case with the Finleys. Rick knew how much the Finleys wanted to spend before he prepared the contract. They weren't going to be surprised.

So when Ed Finley picked up a pen to sign the contract, that wasn't a surprise either.

Chapter 6
Estimating Commercial Work

In the last chapter, we pointed out that remodeling work is fundamentally different from new construction. But in some ways estimating new construction is like estimating repair and remodeling work. In both, you calculate how much labor, material and equipment you need to do the job. You add prices, you summarize costs at the end of the estimate, and then you add an appropriate percentage for overhead and profit.

In this chapter, we introduce the Estimate Take-off Sheet (Figure 6-1) as the first step in the estimating process. The other forms you'll see here are similar to those demonstrated in Chapter 5.

How new construction estimates are different

Here are some of the ways new construction estimates are different from repair and remodeling estimates:

- The job is probably larger.
- There won't be much demolition or removal.
- You'll estimate more structural materials such as concrete and framing.
- You won't have to work around existing construction.
- You shouldn't have to improvise as much because there's less fitting and matching of materials.
- The scope of the job should be well-defined.
- You'll usually receive complete plans and specs.
- The risks tend to be proportionally smaller, there are fewer unknowns, and surprises are less common.

The example estimate in this chapter focuses on concrete. But don't worry. Even though the scope of this estimate is limited, the same estimating procedures still apply, no matter what part of a new construction project you're estimating.

The Commercial Estimate

As in the previous chapter, you'll look over the shoulder of an experienced professional estimator as he works up a bid. You'll follow along step-by-step as the estimator for SCC Contractors creates each part of his estimate. Once you understand the take-off and bidding procedures described here, you should have no trouble adapting the process to any type of building construction. You'll be able to create verifiable, consistently-reliable construction estimates for your company.

Meet Kevin and Chris

Kevin is the owner of SCC Contractors in Mobile, Alabama. He worked as a tradesman, foreman and supervisor in the Mobile area for nearly 20 years before he started SCC Contractors in 1982. From the beginning, SCC specialized in reinforced cast-in-place structural concrete and the associated slabs-on-grade work on commercial projects. Kevin has done well. Annual sales grew quickly to nearly $1 million by the late 1980s.

Kevin's knowledge and high standards helped SCC develop a reputation as a reliable concrete sub. Five of the larger general contractors in the Mobile area usually invite SCC to bid their structural concrete. A typical SCC job takes four to six weeks from start to finish, and carries a bid price of from $50,000 to $120,000.

Kevin, as owner and general manager, is the one who works directly with SCC's general contractor clients. Chris, Kevin's son, does most of the estimating. Kevin reviews all the estimates and decides how much to add for overhead and profit.

When SCC wins a job, Chris becomes the working foreman who supervises the SCC crew as work progresses. Kevin seldom works on a job, but visits the site regularly. He's always available to resolve problems and deal with the architect or project manager. Kevin and Chris also schedule the jobs and work together to order materials for each SCC job.

SCC Office Procedures

Ruthie, the bookkeeper, posts the ledgers, pays bills, handles payroll, answers the phone and takes care of routine correspondence. SCC has a C.P.A. who prepares monthly income statements, quarterly tax returns, and annual financial statements from the information Ruthie supplies. The

C.P.A. also monitors SCC's overhead expenses and advises Kevin on tax and financial matters. Kevin uses the information from his accountant when he decides what percentage to add to each bid for overhead and profit.

Kevin and his wife Dorothy own SCC's office and shop building. It's a 1,200 square foot masonry structure with a large storage yard in an industrial section of Mobile. Kevin stores tools and materials there between jobs. He also uses the shop and yard to repair forms, and to house and maintain company equipment.

SCC Operations on Site

SCC has a permanent two-man construction crew: Chris is the journeyman carpenter and working foreman. His co-worker, Keith, is also a journeyman carpenter. SCC hires a crew of laborers and cement finishers on a job-by-job basis. They can keep this crew busy during most months of the year. The crew is highly experienced in structural concrete forming, placing, and finishing, and works well as a team. They know what SCC expects and usually require little supervision.

Everyone on this crew of "regulars" considers himself a part of the SCC team. They're all included in discussions about the work to be done and how to do it. They respect Chris' ability as a foreman and like to be included in job planning.

Equipment - SCC owns a 55-horsepower Case 580 combination front-end loader and backhoe. This loader-backhoe can handle nearly all the excavation and backfill work on SCC jobs. SCC also owns a 5-ton dump truck and a low-boy trailer they use to haul the pneumatic-tired backhoe to and from each job site. Either Chris or Keith operates the backhoe and drives the truck. SCC also owns a vibrating mechanical screed and several concrete vibrators appropriate for their type of work.

SCC bills part of company-owned-equipment costs to each job. Every year, their accountant calculates the hourly operating cost for SCC equipment. Kevin and Chris use this hourly rate in their job estimates.

The Sunny Hills Job

Early in May, Kevin received a call from Phelps Construction, a general contractor who plans to bid for the first phase of the Sunny Hills Shopping Center west of Mobile. Phelps asked Kevin to prepare a bid on the structural concrete work and have a price ready by noon on May 28. Phelps will submit a sealed bid to the architects by 2 p.m. on that day.

The Fifteen Steps to a Detailed Estimate

Step 1 - Assemble complete information

Step 2 - Scan the drawings

Step 3 - Study the specifications carefully

Step 4 - Visit the site

Step 5 - Contact your subcontractors

Step 6 - Prepare the take-off sheet

Step 7 - Apply unit costs

Step 8 - Extend and total costs

Step 9 - Prepare the project schedule

Step 10 - Re-contact subcontractors

Step 11 - Calculate project overhead

Step 12 - Re-check your math and notes

Step 13 - Verify total cost

Step 14 - Add your profit

Step 15 - Check the bid package

Kevin went immediately to the Phelps office to pick up a set of plans, specs and contract documents. Figure 6-2 shows the concrete plan for the job. Figure 6-5, later in the chapter, shows footing details from that plan, and Figure 6-4 is the part of the job specifications which applies to the structural concrete work. I'll describe those as we come to them.

Kevin noticed immediately that this job was smaller than most of the jobs SCC bids. He also saw that it was the type of work SCC does best and that they would be subcontracting for a general contractor they've worked with many times in the past. Kevin figured the work would fit nicely into a slack period in late July or early August when he had no other jobs scheduled.

Later in the day, Chris looked over the plans and specs with Kevin. They saw that the plans include cast-in-place reinforced concrete continuous wall footings, square column footings, and foundations outside the building perimeter. Chris also noted several concrete curbs adjacent to the building.

Follow a Plan

Back in Chapter 4, I described the fifteen steps in every construction cost estimate. I recommended that you follow these steps in the order listed. The fifteen steps are listed at the top of this page. You'll see that Kevin and Chris follow these steps as they prepare their bid for the Sunny Hills project.

By this time, Chris and Kevin had already done the first three estimating steps. They still hadn't decided whether to bid the project, but that was becoming more and more likely as Chris studied the plans and specs spread out on the table. After an hour, Chris was satisfied that he understood what the project required, and didn't see why SCC couldn't submit a competitive bid.

Visit the Site

The next day, Kevin drove out to the Sunny Hills site. The lot had been cleared and graded and the surveyor's stakes were in place. The only other sign of imminent construction was some soil disturbed when the engineering company bored for samples. While he was on site, Kevin reviewed his site visit checklist to identify potential problems. We showed you one of those back in Chapter 4, Figure 4-2.

Make a sketch of the site and identify potential problems

The only potential problem Kevin saw was a drainage channel that ran by the west end of the property. He guessed that this channel would fill with runoff in wet weather and could flood any excavation on the site. The plans showed a 48-inch storm drain in an easement at the west end of the property. Kevin sketched the site, showing access points from the street, and the street names. On his sketch he wrote *Storm drain* and a big question mark.

Kevin's note reminded him to ask Dave Phelps about drainage problems on the building site. Rain that collects on site can be a problem. SCC's usual procedure is to dig a drain sump at the lowest point of any continuous wall footing. If it rains before the concrete is placed, it's a quick and easy job to pump water out at the drain sump. The extra excavation required is just one dip of the bucket with the backhoe. But if a drainage channel overfills and floods the site, the result is usually construction delays and higher costs.

Back at the office later that day, Kevin described the site to Chris. They agreed that this phase of Sunny Hills was small. But it also looked like a good, clean, quick job that could make money for SCC. Chris promised to begin the quantity take-off by the end of the week.

Start the Take-off

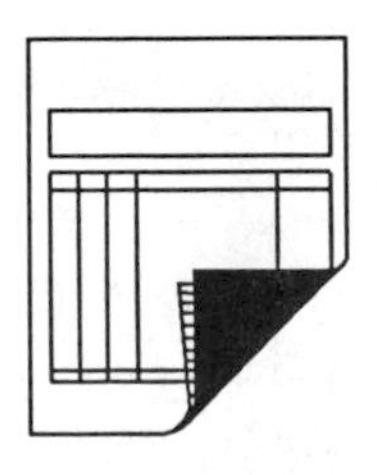

The blank form Chris used as his concrete take-off sheet is shown in Figure 6-1. Completing this form is Step 6 (Prepare the Take-off Sheet) in the fifteen-step estimating process. Figures 6-3 and 6-8 are the actual concrete take-offs Chris prepared. Of course, Chris filled out his estimating forms by hand. In Figures 6-3 and 6-8 we show typed (and somewhat condensed) versions of the estimate Chris prepared. We'll examine this take-off carefully so you understand where each number came from. Notice that the Estimate Take-off Sheet (Figures 6-3 and 6-8) list sizes, not costs.

Estimate Take-off Sheet

Company: ______________________________

Project: ______________________________

Address: ______________________________

Job: ______________________________

CSI Division/Account: ______________________________

Estimator: ______________ Date: ______________

Checked by: ______________ Date: ______________

Estimate due: ______________

Estimate # ______________

Drawing reference: ______________

Description	Q=Qty	L=Length	W=Width	D=Depth	T=Thickness	H=Height	Calculation	Total/Unit

(Carry description and totals forward to Estimate Detail Sheet)

Figure 6-1 Estimate take-off sheet

Figure 6-9 (Estimate Detail Form) is the cost estimate. This form lists costs and quantities, but not sizes. Completing Figure 6-9 is Step 7 (Apply Unit Costs) and Step 8 (Extend and Total Costs) in the fifteen-step estimating process I discussed in Chapter 4.

Costs on the Estimate Detail Form are summarized on Figure 6-10, the Estimate Summary.

The Audit Trail

Chris was careful to create an *audit trail* for each cost in the Estimate Summary. Anyone who questions a cost on the Estimate Summary can refer back to the Estimate Detail to check quantities or costs. To find dimensions and sizes used to compute quantities, go back one step further to the Estimate Take-off Sheet. To verify information on the Estimate Take-off Sheet, take another step back to the plans and specs. That's called the audit trail. Creating that trail is an essential part of every professional estimator's trade.

Good estimates are verifiable

Good estimates, like scientific experiments, are verifiable: Each step in the process can be checked by anyone who understands the procedure. If there's a mistake somewhere in the estimate, it's likely to be obvious to another estimator. Verifiable estimates are easy to check. That usually makes them much more accurate.

Chris writes notes on his take-off sheets that make it easy for Kevin (or anyone else who wants to check this estimate) to understand where the numbers came from. That's good practice for any estimator. Anything that isn't obvious should be explained by a note that shows how you arrived at the quantities and costs.

The Take-off Form

Leave room to list dimensions

Chris likes this Estimate Take-off Sheet format because there's room to list each dimension shown on the plans. Notice the column headings on the Estimate Take-off Sheet:

Description	T=Thickness
Q=Quantity	H=Height
L=Length	Calculation
W=Width	Total/Unit
D=Depth	

You compute the quantities on the Estimate Take-off Sheet, then transfer them to the Estimate Detail Form, Figure 6-9. There's a line at the bottom of our Estimate Take-off Sheet which reminds you to do that.

In many cases, you list dimensions on the Estimate Take-off Sheet one time but use them many times. For example, look under the heading *Continuous wall footings*. Chris used the same dimensions to calculate cubic yards of earth to be excavated, linear feet of form lumber required, cubic yards of concrete, pounds of reinforcing steel, and cubic yards of backfill required. Those dimensions go on the BBD (Basic Building Data) list we described in Chapter 4.

Cost Records

Keep track of actual cost

Kevin and Chris always keep track of actual costs on their jobs. They can predict fairly accurately the cost of work they do often. For example, they have good cost records per cubic yard for foundations and curbs and costs per square foot for slabs on grade. If you don't have good cost records for repetitive tasks, I recommend that you start developing unit price records on your next job. Chapter 8 has more information about recording actual costs.

What if you don't have records of your actual unit costs? Refer to published estimating references such as those listed in the order form at the end of this manual. The CD-ROM inside the back cover of this book contains an estimating program for *Windows*, and all the cost data in the *National Construction Estimator*, the *National Repair & Remodeling Estimator*, the *National Plumbing & HVAC Estimator*, the *National Electrical Estimator*, the *National Painting Cost Estimator* and the *National Renovation & Insurance Repair Estimator.* Of course, you can use the printed versions of these books without the CD-ROM. But the software will save you time, and prevent most math errors.

But remember, no published reference is based on materials from *your* supply yards, installed by *your* crews on *your* type of project. But if published cost reference data is all you've got, you'll have to use it.

How to Take Off Quantities

Look at Figure 6-2, which shows the foundation plan Chris used to find most of the dimensions for the Sunny Hills job. This is just one page from the plan set. Chris also used elevation views and section details to find some dimensions for his estimate.

Work very systematically when you do a material take-off. In our example estimate, Chris began at the upper left of Figure 6-2 and moved clockwise around the drawing. When you do a take-off, put a check mark over each item on the drawing as you write it down. The check mark shows that you've included that item in the take-off. This makes checking your take-off very easy. For example, Chris used a green marker to shade each footing on the plan after he listed the dimensions on the Estimate Take-off Sheet. He used the same green color code for footing dimensions on the Estimate Take-off Sheet itself.

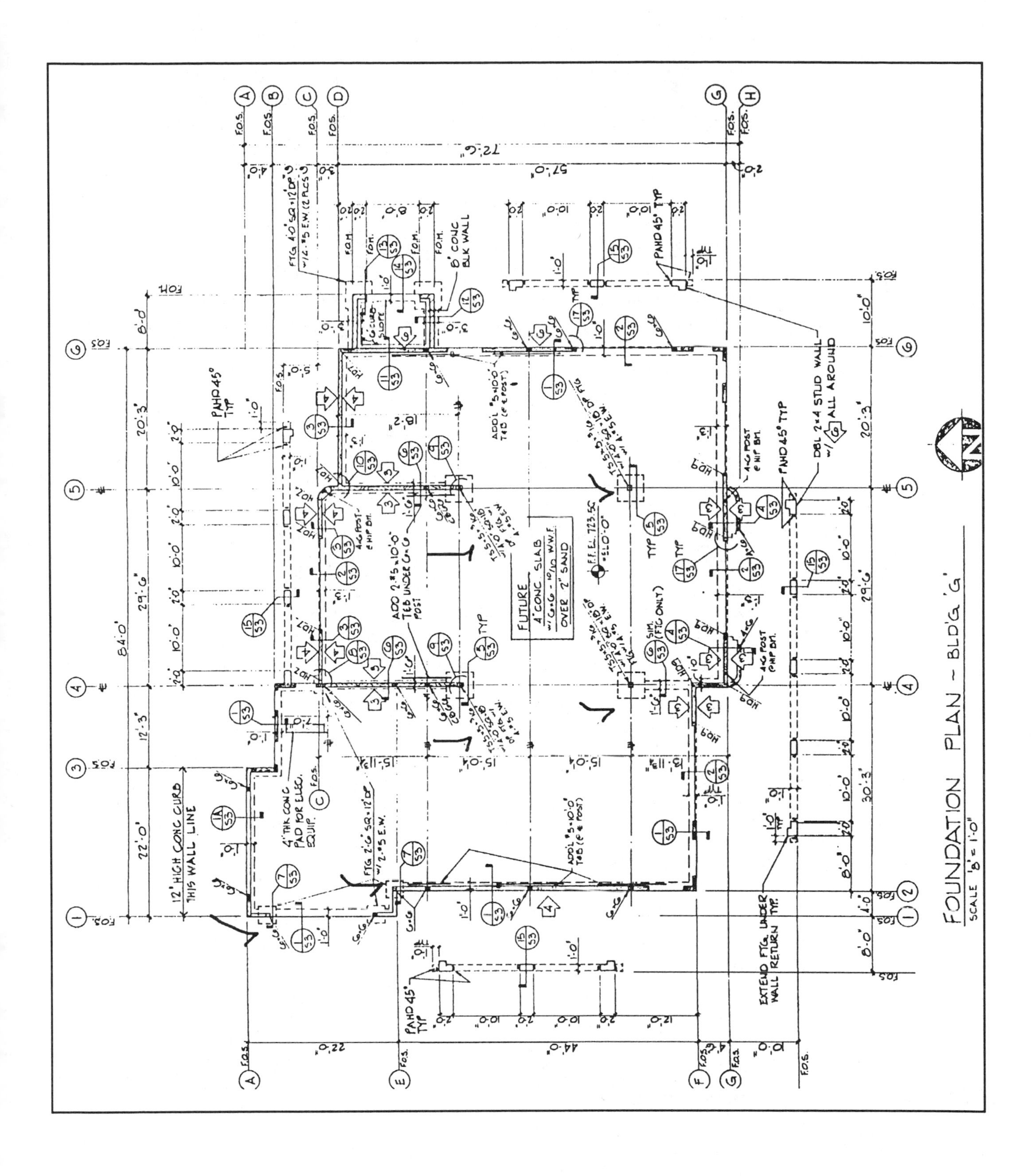

Figure 6-2 Concrete plan

Work with one category at a time

Always work with a single category of the take-off (such as the footings) until you've accounted for everything in that category. If someone else has to finish the take-off, or if you're interrupted before you finish, it will be easy to pick up where you left off.

Start at the Top

Use the top section of the Estimate Take-off Sheet to identify your job, as Chris did on Figure 6-3. Write the Masterformat division number on your take-off sheet also. In our example, that's *3.0/Concrete*. Under "Drawing reference," write the page number of the plan document where the information for each take-off sheet came from.

Do the take-off in the same order you'll build the job. That makes it less likely that you'll overlook something. It also makes it easier to check the estimate. In this case, the first job was to lay out the foundation. That's the first item under the *Description* column in Figure 6-3.

Notice the line that says, *control points and site prep by others NOT included in SCC Scope of Work.* Survey markers, also known as control points or bench marks, are usually set by the general contractor's surveyor. These marks identify both horizontal position and elevation. The foundation plan shows where the bench mark control points are located. One of these bench mark control points is in the center of Figure 6-2 just below the note about the future 4-inch concrete slab. It's a circle with two opposite quarters shaded. It refers to the finished floor elevation. These points are usually marked on the ground with a permanent monument, such as a brass tack driven into a concrete curb. There are other control points for this job, but they're not shown on the drawings we are using.

"Do the take-off in the same order you'll build the job. That makes it less likely that you'll overlook something."

Always write down any assumptions you make about what's included, or not included, in your part of the job. Chris assumes that the general contractor will provide SCC with survey markers and a building pad ready for the foundation work to begin.

Working from this control point, Chris and his crew will lay out the foundation from measurements shown on the plans. They'll use the elevation information to determine the depth of footings and the finished height of the foundations.

Estimate Take-off Sheet

Company: SCC Contractors
Project: Sunny Hills Shopping Center Building #1
Address: State Rt. 70 & Old Church Road
Job: Dave Phelps General Contractor Job #93-05-12
CSI Division/Account: 3.0/Concrete

Estimator: Chris Riley **Date:** May 17, 1993
Checked by: Kevin Riley **Date:** May 21, 1993
Estimate due: May 28, 1993
Estimate # 06-1
Drawing reference: (Figure 6-2)

Description	Q=Qty	L=Length	W=Width	D=Depth	T=Thickness	H=Height	Calculation	Total/Unit
Lay out foundations. Set batterboards, chalk mark excavations	Note: control points and site prep by others NOT included in SCC Scope of Work							
Exterior corners	14	--	--	--	--	--	Q x 1	14 Ea
Footings (calculate as 4 corners each)	12	--	--	--	--	--	Q x 4	48 Ea
							Layout points total	62 Ea
Excavation Continuous wall footings (dimensions in feet) Building perimeter & inside building	Note: All concrete extends 6" above top of excavation. Add 33% to excavation for same. (6"=1/3 of 18") Note: Formed haunches 6" deep required both sides. Use 2 x 6 each side, trim to fit.							
22'+4'+12.25'+57'+4.5'+30.25'+44'+4'+22'		200.0	1.0	1.5	--	--	L x W x D/27	11.11 CY
29.5'+3'+20.25'+20.25'+29.5'		102.5	1.25	1.5	--	--	L x W x D/27	7.12 CY
18.2'+18.2'+6.5'		42.9	1.5	1.5	--	--	L x W x D/27	3.58 CY
Outside building								
39'+8'+28'+52'+28'		155.0	1.0	1.5	--	--	L x W x D/27	8.61 CY
8'+8'		16.0	3.0	1.5	--	--	L x W x D/27	2.67 CY
Square footings (dimensions in feet)								
Inside building	4	2.5	2.5	1.5	--	--	Q x L x W x D/27	1.39 CY
Inside building	4	4.0	4.0	1.5	--	--	Q x L x W x D/27	3.56 CY
Outside building	2	4.0	4.0	1.5	--	--	Q x L x W x D/27	1.78 CY
Total excavation					**Excavation, including 5% for overbreak**		All CY + 5%	**42 CY**
Concrete **Total concrete**					**Concrete, including 6" extra height**		Excav + 33%	56 CY
Reinforcing bars Reinforcing based on total length in footings plus 10% for waste and laps								
3' #3 dowels @ 1.33' OC wall footings	LF	1281	--	--	--	--	L x .38 Lbs	487 Lbs
3' #4 dowels @ 2.00' OC wall footings	LF	852	--	--	--	--	L x .38 Lbs	324 Lbs
6.5' #4 ties @ .58' OC for 1.25" wide ftgs.	LF	1264	--	--	--	--	L x .38 Lbs	480 Lbs
	#3 and #4 reinforcing bars including 10% waste & laps					**Total**	**All Above**	**1291 Lbs**
One #5 T&B for 1.0' wide footings	LF	440	--	--	--	--	L x 1.04 Lbs.	458 Lbs
18 pieces #5 at 10' for extra bars at columns	LF	198	--	--	--	--	L x 1.04 Lbs.	206 Lbs
Two #5 E.W. for 2.5' square footings	LF	22	--	--	--	--	L x 1.04 Lbs.	23 Lbs
Four #5 E.W. for 4' square footings	LF	211	--	--	--	--	L x 1.04 Lbs.	220 Lbs
Two #6 T&B for 1.5' and 3.0' wide footings	LF	259	--	--	--	--	L x 1.50 Lbs.	389 Lbs
Three #7 T&B for 1.25' wide footings	LF	777	--	--	--	--	L x 2.04 Lbs.	1586 Lbs
	#5, #6 & #7 reinforcing bars including 10% waste & laps					**Total**	**All Above**	**2882 Lbs**

(Carry Description and Totals forward to Estimate Detail Sheet)

Take-off sheet 1 of 2

Figure 6-3 Take-off sheet for excavation, concrete and rebar

The Foundation Layout

Chris began by counting the corners. Note on Figure 6-3 that Chris counted 14 exterior corners. For estimating purposes, Chris considers the number of corners to be a good measure of the time required to lay out the foundation. The more corners, the longer the layout will take.

Chris also counted the number of footings that aren't building corners. He found twelve of those footings and wrote that number in the quantity column on Figure 6-3. Chris figured that it will take about as much time to lay out each of these footings as to lay out four corners. This system worked for Chris in the past, so he decided to use it again here. You can see the number *12* in the second line of the *Q=Qty* column in Figure 6-3.

Of course, there are other ways to estimate the time required for foundation layout. For example, you could estimate layout time based on square feet of slab or linear feet of foundation. But Chris recognized that the number of corners on this job made layout more complex. So he estimated layout by the corner.

Footing Excavation

After the layout is done, the next step is to excavate for the footings. Chris noticed right away that he could kill two birds with one stone on this take-off. The volume of soil excavated for the footings will be nearly the same as the volume of concrete needed to form the footings.

The specifications (Figure 6-4) require a footing depth of 18 inches (1.5 feet) below the adjacent grade. Footings for this building extend 6 inches above grade, as shown in detail 1A in Figure 6-5. Footing depth below grade plus 6 inches is the usual footing height.

First, Chris calculated the volume of the footing trench. Then he used a simple adjustment to convert soil volume to concrete volume for the footings. Chris wrote the heading *Excavation* on his Estimate Take-off Sheet.

Plans are like a map

Notice the north arrow at the bottom of the plan (Figure 6-2). Turn this north arrow so it's pointing away from you. Across the top (north side) of the plan you'll find lines numbered 1 and 3 to 6. On the bottom (south side) the lines are numbered 1, 2, and 4 to 6. These are called *column lines*. Down the right (east) side of the plan you'll see *row lines* lettered from A to H (E and F are on the west side). These column and row lines work like the grid lines on a map, and provide a handy way to identify key points on the plan. Chris started his footing take-off at point A1 (where row line A intersects column line 1) at the top left of the plan. From there he worked clockwise around the building perimeter.

The length of the first footing is 22 feet. Note the numbers 22′-0″ between column lines 1 and 3 at the top of Figure 6-2. But what's the depth and width of this trench? The plan shows the width as 1 foot. You see 1′-0″ drawn across the footing, indicating the width. Now, all we need is the depth to calculate the volume. Chris had to go to a detail drawing to find the trench depth.

General

1. All footings shall be founded a minimum of 18" below the lowest adjacent grade and a minimum width of 12 inches. Allowable soil bearing pressure for design is $Q = 1750 + 350\ (W-1) + 550\ (P-15) \lesssim 4500$ P.S.F.
2. Footing elevations shown are for bidding purposes only and are assumed to be in suitable bearing materials. The actual adequacy of the bearing material shall be determined by the soils engineer prior to placing of reinforcing or pouring of concrete and footing elevations shall be lowered as directed by this representative, if necessary.

Reinforced Concrete (Regular weight 145 P.C.F.)

1. Concrete strength: The concrete strength shown in the following table is the minimum compressive strength at 28 days: The aggregate shown is the maximum size: and the slump shown is the maximum in inches.

Item of construction	Strength (PSI)	Aggregate (inches)	Slump (inches)
Foundations	2,000	1-1/2	4
Slab-on-grade	2,000	1	3

2. Reinforcing steel shall conform to ASTM A 615: No. 5 & larger, Grade 60; No. 4 & smaller, Grade 40.
3. Welded wire fabric shall conform to ASTM A-185, flat sheets.
4. No pipes or ducts shall be placed in concrete columns, walls or slabs unless specifically detailed.
5. Reinforcing, anchor bolts and all other embedded items shall be securely held in position prior to placing concrete.
6. Walls and columns shall be doweled from supports with bars of the same size and spacing.
7. Splice continuous reinforcing with 40 diameter or 1′- 6″ minimum lap.
8. Typical concrete coverage of reinforcing:

Concrete cast against earth — **3"**
Exposed to earth or weather - larger than #5 — **2"**
- #5 and smaller — **1½"**
Unexposed - columns, beams and girders — **1½"**
- slabs and joists — **¾"**

9. Spacer ties: Furnish #3 ties at 24 inches in all beams and footings, unless otherwise shown in details.
10. See architectural, mechanical and electrical drawings for locations of pipes, ducts, vents, special inserts, moulds and other embedded non-structural items.
11. Reinforcing steel shop drawings shall be reviewed by the architect before fabrication.

Figure 6-4 Job specifications

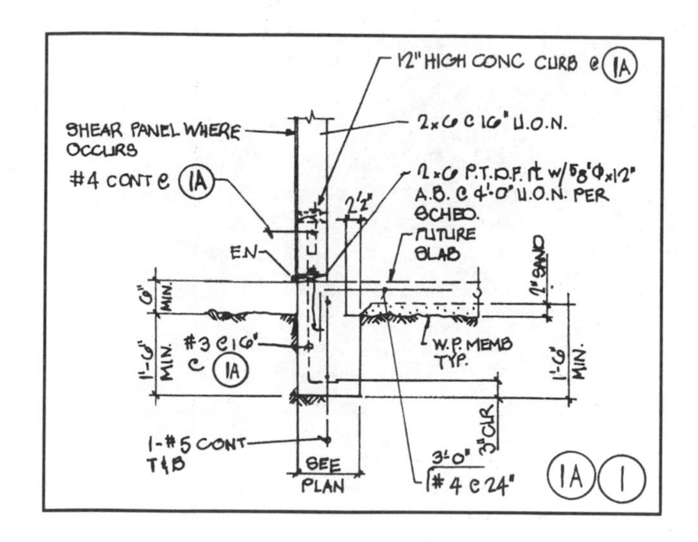

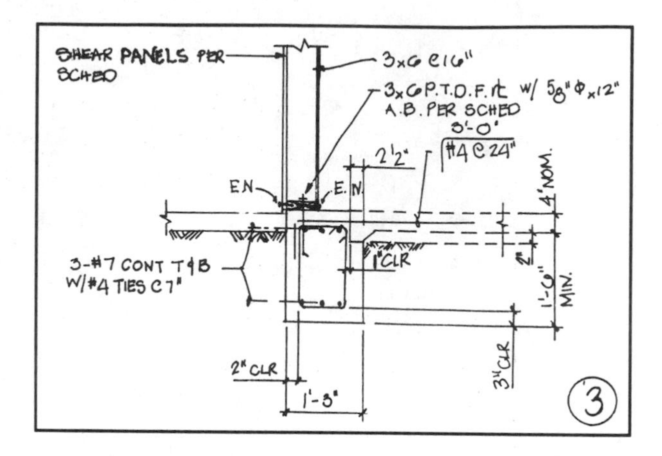

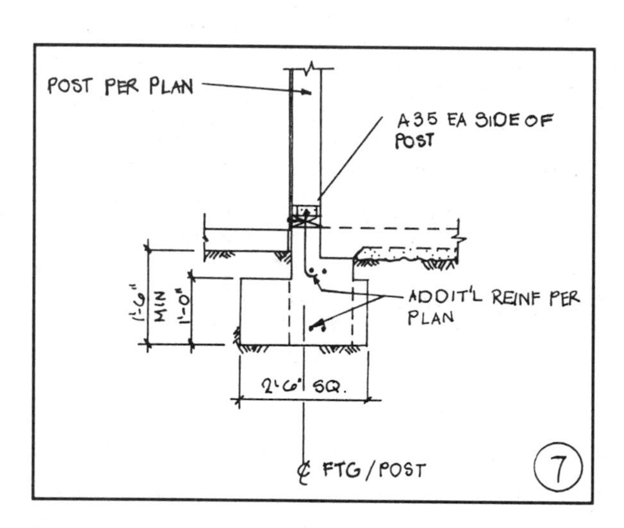

Figure 6-5 Detail drawings

Notice the circle with *1A* and *S3* next to the 22-foot footing. That means more information on that footing is found on plan sheet S3 and in detail drawing 1A. Figure 6-5 shows several details from plan sheet S3. Find detail 1A and you'll see that the minimum footing depth below the adjacent grade is labeled 1′-6″. That's equal to 1.5 feet. Chris made 1 foot wide by 1.5 feet deep his first category for footing excavation in the take-off section under "Continuous wall footings." These numbers appear in the *W=Width* and *D=Depth* columns in the take-off section under "Continuous wall footings." This line will include all the footings of this size which fall on or inside the building perimeter.

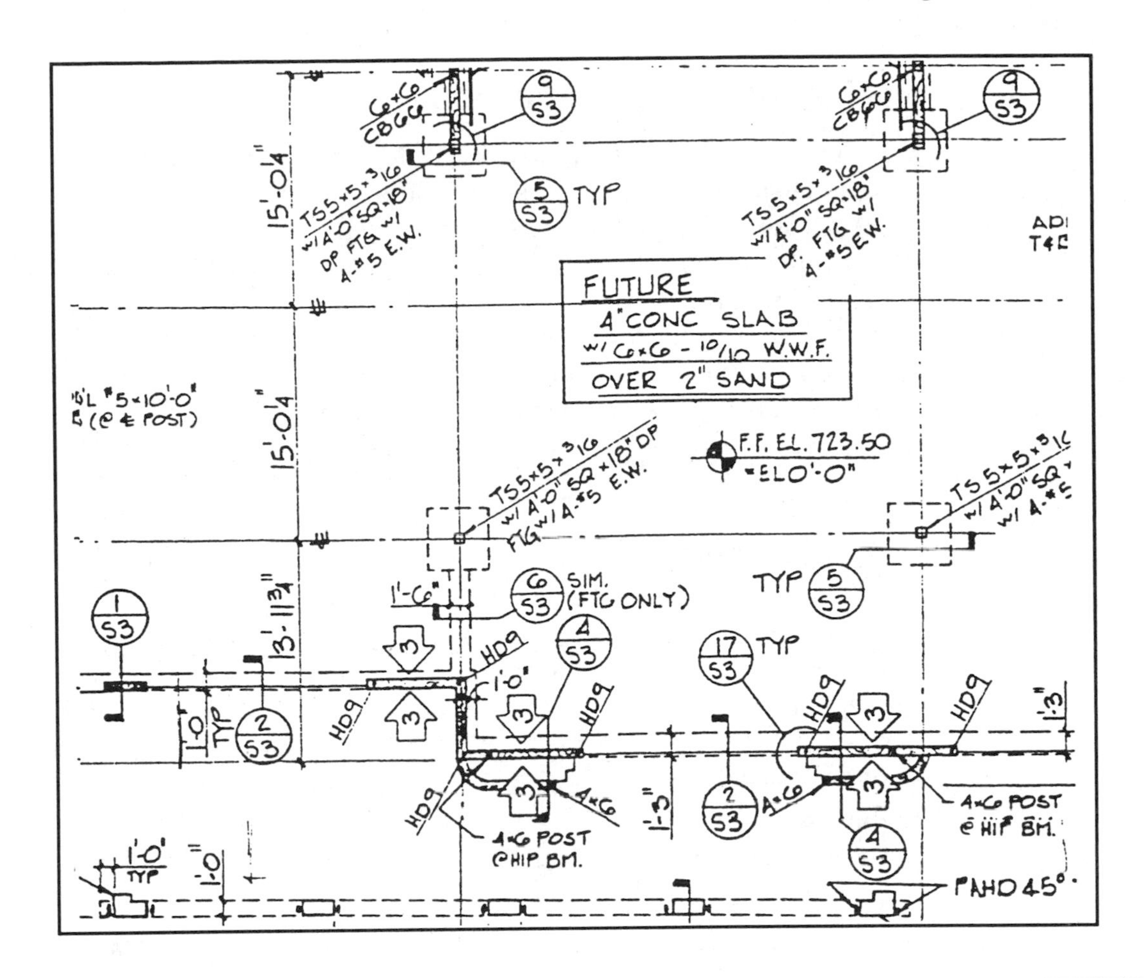

Figure 6-6 Section of concrete plan

Chris wrote 22′ at the left of this line on Figure 6-3 because the first footing trench in this size category was 22 feet long. Then Chris marked that footing line with his colored pencil to show that he had listed it on his take-off form.

Working clockwise, you see (back on Figure 6-2) that the next footing trench that measures 1 foot wide by 1.5 feet deep is between A3 and B3 on the plan. It's a short footing, only 4 feet long. (Notice 4′-0″ written between row lines A and B at the right margin of Figure 6-2.) This footing is either a detail 1A footing or a detail 1 footing. Chris wasn't sure which it was, because the detail reference shown for the 22 foot-long footing on the left shows Detail 1A, and the 12 foot 3 inch footing on the right shows Detail 1. But it doesn't matter, because both are 1.5 feet deep. So Chris listed this short trench as the second on his take-off form, Figure 6-3. He wrote +4′ on the line for footings 1 foot wide by 1.5 feet deep.

If you continue around the plan, you'll see a section of footing for each of the numbers on that line of the Estimate Take-off Sheet. Always convert feet and inches to feet and hundredths of a foot. For example, the next footing section is 12′-3″ on the plan (between columns 3 and 4) but it appears on the take-off as +12.25′.

Taking Off the Larger Footings

Once he had all the 1-foot-wide by 1.5-feet-deep building perimeter footings listed on his take-off sheet, Chris looked for the 1-foot, 3-inch-wide and 1.5-feet-deep footings. That's the second category under "Excavation" on Figure 6-3. The footing that runs from C4 to C5 is 29 feet 6 inches long. According to detail 3 in Figure 6-5, it's 1 foot 3 inches wide and a minimum of 1 foot 6 inches deep. Chris entered 29.5′ on the second line under continuous footings. He shaded the 1′-3″ x 1′-6″ footings with a different color on the drawing and on his Estimate Take-off Sheet.

He continued listing the perimeter footings by category until he had checked all of them off with his colored pencil.

Then Chris listed the interior continuous wall footings. One of them is a 1.5-feet-wide by 1.5-feet-deep footing on column line 4 at the south wall, as shown in Figure 6-2, and the enlarged drawing in Figure 6-6. Chris had to calculate the length of that footing. Note the 13′-11¾″ dimension next to the square footing. From this, Chris subtracted the total of 4 feet 6 inches (the setback between rows F and G), plus 1 foot for the width of the south wall footing, plus 2 feet for half the square footing. This gave him 6 feet 6 inches, so he entered +6.5′ as the last dimension for the 1.5-feet-wide by 1.5-feet-deep footings. He shaded these footings with colored pencil as he listed them on the take-off sheet.

The next two lines on the take-off are for the footings outside the building.

Square Footings

Begin locating the square footings at the upper left corner of Figure 6-7. Near the bottom and to the right of column line 1, you see *FTG 2′6″ SQ x 12″ DP*, with arrows pointing to four locations. Those are on the first line for square footings on the take-off. There's more information for these footings on detail 7 of drawing S3, our Figure 6-5. That's where the depth of 1.5 feet came from.

On Figure 6-6, above the note about the future concrete slab, there are two 4-foot-square footings. Two more of these square footings are below the note about the slab. Those are on the second line of Figure 6-3 under "Square footings."

Two more square footings are located outside the northeast corner of the building 8 feet east of column line 6. These footings are also 4.0 feet square. Those are on the take-off sheet with a note that they're outside the building.

As Chris listed the footings, he wrote the dimensions for each category in the columns headed *L=Length, W=Width, and D-Depth* on Figure 6-3. In the *Calculation* column he wrote the formula for calculating the

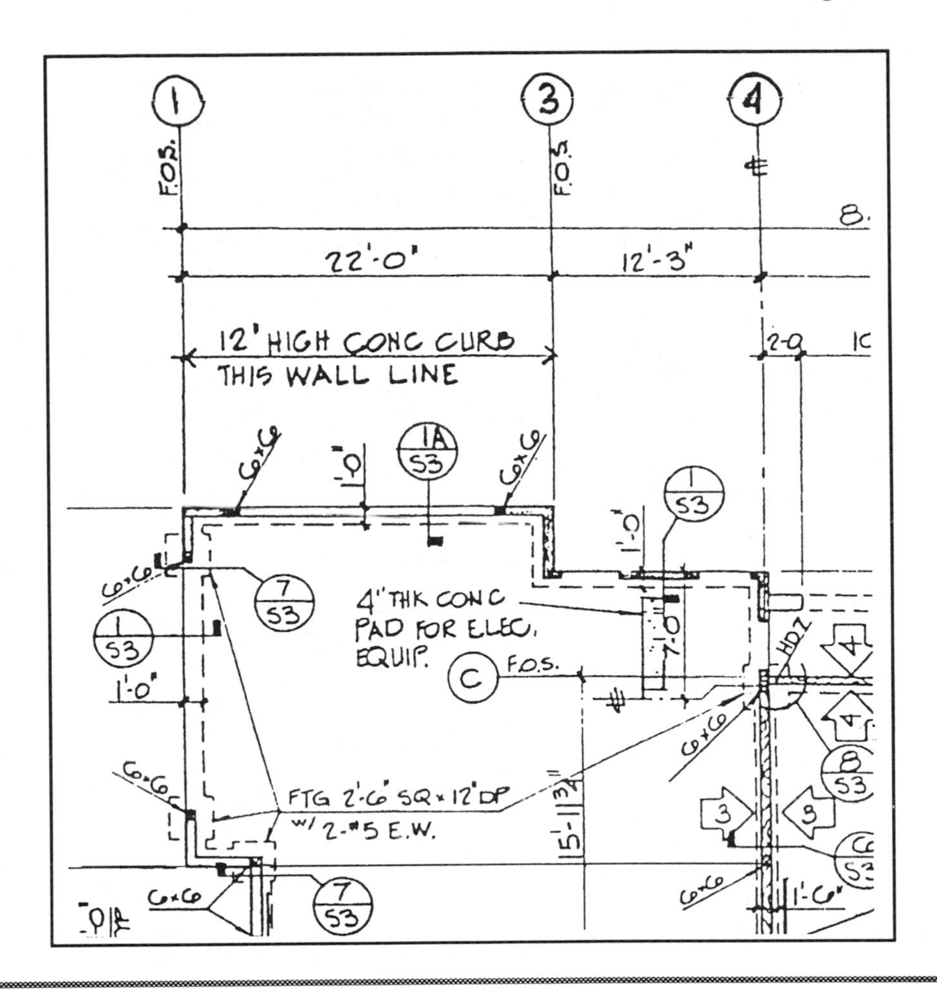

Figure 6-7 Section of concrete plan

excavation volume of each type of footing. The formula, *QxLxWxD/27* is shorthand for "volume in cubic yards equals quantity times length times width times depth in feet and divided by 27 (the number of cubic feet in a cubic yard)."

The Excavation Quantity

The *Total/Unit* column contains the result of the calculation for each line. For example, on the first line of continuous wall footings, 200.0 times 1.0 times 1.5 divided by 27 is 11.11 cubic yards of soil to be excavated.

Add the numbers in the *Total/Unit* column, and you see that the total excavation volume is 39.82 cubic yards. Chris added 5 percent to this for over-excavation, and rounded the answer to an even 42 CY.

The Concrete Quantity

Once you have the excavation quantity, it's easy to turn that into concrete quantity. The detail drawings show that the foundation concrete extends 6 inches above the adjacent grade. Except for this extra 6 inches, the concrete quantity is the same as the excavation quantity. Since all the footings were 1.5 feet deep, add 1/3 (or 33 percent) to allow for the extra 6 inches of concrete (1/3 of 1.5 feet equals 0.5 feet, or 6 inches).

Notice the line for "Concrete" on Figure 6-3. Chris noted that the concrete quantity was based on the excavation quantity plus 33 percent. Then he wrote 56 CY in the total column because 42 CY plus 33 percent is 55.86 cubic yards.

Reinforcing Bars

The next heading on Chris' take-off sheet is for "Reinforcing bars." Notice that Chris will allow 10 percent for waste and laps. The minimum lap for rebars is set in note 7 of the specs, shown in Figure 6-4. Chris based the 10 percent waste allowance on past experience with laps of 40 diameters, or a 1-foot 6-inch minimum.

Rebar is sold by weight, but Chris estimates it by the linear foot. To convert from linear feet to weight, multiply the length of each size bar (in feet) by the weight per foot. Of course, weight varies with the size of the bar. Rebar numbers indicate the size of the bar in eighths of an inch. For example, a #4 bar is 4 times 1/8, or 1/2 inch in diameter. A #6 bar is 6/8 (3/4 inch) in diameter. You can see the weight per foot for each size bar in the *Calculation* column of the Estimate Take-off Sheet.

From item 2 of the specifications (Figure 6-4) under Reinforced Concrete, you see that #4 and smaller rebar is specified as Grade 40. Number 5 and larger is Grade 60. Chris listed these grades in separate categories on the take-off sheet for two reasons:

1) There could be a difference in price
2) The separate listing confirms that Chris followed the specs

Taking Off Dowels for Masonry Walls

Refer to the drawings to find the location, quantity and length of the rebars. A typical callout for reinforcing bars is shown for the 4-foot square 18-inch-deep footing at the end of the short continuous footing on Figure 6-6. Above the arrow, you see "TS 5 x 5 x 3/16." That's for the steel column that's not part of SCC's scope of work. But below the arrow, following the footing dimensions, it says, "4 - #5 E.W." which means *four #5 bars each way*. See also in item 6 of the specifications (Figure 6-4) that wall footings require dowels. These are short lengths of rebar inserted in the footings to tie the footing to the wall above and to the slab.

Three types of reinforcing bars

Detail 1A in Figure 6-5 shows three types of reinforcing bars:

1) #3 bars (dowels) for the wall at 16 inches OC (on center)
2) #4 bars (dowels) for the floor slab at 24 inches OC
3) Continuous #5 bar (*CONT* on detail 1A) at the top and bottom (*T&B* on detail 1A) for the footing

Start with the #3 dowels. The length of these dowels is the footing depth of 1.5 feet, plus the 6 inches (0.5 foot) of concrete above grade, plus 1 foot of projection above the top of the footing, plus 3 inches (0.25 foot) for the hook, less 3 inches (0.25 foot) for clearance between the bottom of the hook and the bottom of the footing. That came to 3 feet per dowel (1.5 + 0.5 + 1 + 0.25 - 0.25). Chris wrote:

3′ #3 dowels @ 1.33′ OC wall footings

as the first entry under reinforcing steel.

To find the quantity of #3 dowels, add the lengths for all continuous footings (200′ + 102.5′ + 42.9′ + 155.0′ + 16.0′). That's 516.4 linear feet of footing. Add 10 percent for laps to get 568.04 linear feet. Then divide by the 1.33 center-to-center spacing. The result is 427 dowels.

Since each dowel has to be 3 feet long, you multiply 427 by 3 to get 1,281 linear feet of #3 rebar. That number is in the *L=Length* column. In the calculation column, you multiply 1,281 linear feet by 0.38 (the bar weight in pounds per linear foot) to get 487 pounds of #3 bar.

Chris figured rebars for the other footings the same way. He added the footing lengths for each type bar and multiplied the length of the footings by the number of bars to find the total bar length. Bar length times the pounds per linear foot gave him the total weight for each type of bar on his take-off sheet. Then he summarized the total weight of #4 and smaller bars and #5 and larger bars. That completed the rebar take-off.

The Formwork Take-off

Chris used the footing lengths he already computed to find how many forms he'd need. Since all footings would be cast against the earth, he didn't need forms for most of the footings. But note in detail drawing 1, in Figure 6-5, that the top of the wall footing extends 6 inches above the top of the excavation. Chris wrote this opposite the heading "Excavation" on Figure 6-3.

Figure 6-8 is page 2 of the Estimate Take-off Sheet. The top of this sheet is filled in the same as page 1. The first line under *Description* shows *Forms: Haunches and blockouts, both sides wall footings*. The next two lines describe the material Chris planned to use for those forms, two 2 x 6s on each side of the footing. This note creates an audit trail. It leaves no doubt about how the forming was estimated.

The word *haunches* describes the ledges formed in the foundation to support masonry block that will be placed later. Chris will also use these haunch forms as his rebar stringers, which hold the bars in position until the concrete has set. He and Keith will place the haunch forms and set stakes to hold them. They'll tie rebars to these forms with tie wire. (Cost for the ties is built into the allowance for laps and waste in the rebar.)

Estimate Take-off Sheet

Company: SCC Contractors
Project: Sunny Hills Shopping Center Building #1
Address: State Rt. 70 & Old Church Road
Job: Dave Phelps General Contractor Job #93-05-12
CSI Division/Account: 3.0/Concrete

Estimator: Chris Riley **Date:** May 17, 1993
Checked by: Kevin Riley **Date:** May 21, 1993
Estimate due: May 28, 1993
Estimate # 06-1
Drawing reference: (Figure 6-2)

Description	Q=Qty	L=Length	W=Width	D=Depth	T=Thickness	H=Height	Calculation	Total/Unit
Forms: Haunches and blockouts, both sides wall footings Allow double 2 x 6, trim to fit depth per drawings See Take-off Sheet 1, total length all wall footings, times 2 (for both sides) with 2 boards each side (deep)								
200'+102.5'+42.9+155'+16'=516.4'	2	516.4	--	2	--	--	QxLxD	2066 LF
Include strip and clean all wall forms on Estimate Detail Sheet								
Rebar stringers								62 Ea
Wall forms. Included with haunch forms								
Square footings, 4 pieces per footing, add 2' to L								
2.5' square, total 4	16	4.5	Use scrap from haunches					included above
4.0' square, total 6	24	6.0	Use scrap from haunches					
Continuous footings outside the building:	1	171	Use scrap from haunches					
	Total forms including 10% waste, Haunches (2 x 6)							2273 LF
Backfill and compaction	See Take-off Sheet 1. Excavation=42 CY and Concrete=56 CY							
Around foundations, minimum compaction, wheel roll								
Backfill (42 CY) less 75% of ready-mix (56 CY concrete) = 0 CY; Allow 4 CY				Total backfill and compaction			Allowance	4 CY
Stockpile excess on site for use by others (Excavation 42 CY, less backfill 4 CY)				Spread excess excavation on site			42-4	38 CY
Miscellaneous concrete, other than foundations								
Curbs, straight, formed, 2000 psi ready-mix concrete placed direct from chute of ready-mix truck								
Use SCC standard assembly price for cast-in-place concrete curbs								
6" wide x 12" high	1	22			Total 6" x 12" curb		Q x L	22 LF
6" wide x 6" high	1	28			Total 6" x 6" curb		Q x L	28 LF
Alternate for future slab-on-grade								
Concrete slab-on-grade, with 6 x 6 - 10 x 10 welded wire fabric reinforcing over 2" sand								
Use SCC standard assembly price, typical 4" slab	5,000	SF		Alternate for future 4" thick slab-on-grade				5000 SF
Note: add for pumping of concrete if slab placed after masonry walls are in place								
Note: Building slab and electrical pad NOT included in SCC scope of work. See alternate for future 4" thick concrete slab								

(Carry description and Totals forward to Estimate Detail Sheet)

Take-off sheet 2 of 2

Figure 6-8 Take-off sheet for forms, stringers, backfill and curbs

Chris knows that the total length of continuous footing form is 516.4 feet. He entered that figure in the *L=Length* column. The *Q=Quantity* column shows 2, because these forms will be set on both sides of the footings.

The *D=Depth* column shows 2 because two 2 x 6 boards will be required on each side. That's because a 2 x 6 is only 5½-inches wide. The plans call for a 6-inch minimum depth for the haunches. One 2 x 6 won't be enough, so Chris had to estimate two 2 x 6 boards on each side of the footing. He could have used 2 x 4s in this case, but 2 x 6 form lumber is more practical to salvage for future jobs, and he knew they already had some in their storage yard.

The length of 2 x 6 required is the footing length (516.4) times the number of sides (2) times the depth (2). The answer is 2065.6. That's rounded up to 2066 in the *Total/Unit* column.

Chris didn't list any extra material on his take-off form for the rebar stringers on the square footings or footings outside the building. He expects to use scrap pieces left over from the haunch forms for this purpose. He wrote that on the take-off sheet. He also added 10 percent for waste, which will also allow for the haunches.

Backfill and Compaction

After the concrete is poured, and the forms are stripped and cleaned, the next step is to backfill and compact around the foundations. Since that's the construction sequence, that's the order you'd follow to estimate the job.

Chris carried forward the quantities calculated for excavation and concrete from Figure 6-3. You remember that concrete volume was estimated at 56 cubic yards, based on the total excavation volume of 42 cubic yards.

On most jobs Chris figures that the volume of backfill will be the volume of soil excavated less the volume of concrete poured below grade. In this case, 75 percent of the concrete will be poured below. Why 75 percent? Remember that footings are 1.5 feet deep and extend 0.5 foot above ground level. The below-ground portion is 75 percent of the total concrete. Chris did the calculations: 56 cubic yards of concrete times 0.75 is 42 cubic yards. Subtracting 42 cubic yards from the total excavated (42 cubic yards) yields zero cubic yards for backfill.

Chris figured the volume the long way. You probably recognized immediately that there would be little or no backfill on this job. Still, Chris had to estimate the cost to dispose of excavated soil. He allowed 4 cubic yards for any backfill that might be needed, and made a note that the excess (38 cubic yards) would be stockpiled on site for use by others.

The specifications didn't include a requirement for compaction, so Chris made a note that compaction was "minimum." He planned to compact any backfill by rolling it with SCC's rubber-tired Case backhoe.

Slab and Miscellaneous Concrete

Chris didn't bother to figure the time and materials needed to form, pour and finish the straight curbs and the 4-inch slab. SCC has good cost data on this type of work. Chris simply noted on his take-off sheet that he was going to use SCC standard assembly pricing for these items.

Check totals against previous jobs

The reason Chris did a detailed take-off on the foundations is because foundation costs vary from job to job. Chris knows that SCC's cost for small footings cast against soil is usually $135 to $150 per cubic yard, including excavation, forming, reinforcing, and backfill. But foundation costs can be a lot more, and occasionally will be less. The only way to be sure of the cost is to do a detailed take-off. Still, Chris will check his totals against cost records on previous similar jobs. If the estimated costs on the Sunny Hills job are much higher or lower than usual, he'll check his estimate for an error.

Applying Unit Costs

When he finished the Estimate Take-off Sheet, Chris was ready to transfer quantities to the Estimate Detail Form, Figure 6-9. That's Step 7 in the 15-step estimating process. The line below the heading on Figure 6-9: *(Totals and descriptions carried forward from Estimate Take-off Sheet)* shows the audit trail. Anyone who reviews this page of the estimate won't have to wonder where the information came from.

Again, fill in the heading on the Estimate Detail Form. Make a note on this form about anything not included in the scope of work. Some of that already appears on the Estimate Take-off Sheet, but it's a good idea to write it here also. Notice that Chris also indicated that the estimate is based on a two-man crew, and explained how he arrived at the cost per manhour for the labor. Chris knows that Kevin will check the estimate before he submits SCC's bid. These notes will eliminate the most obvious questions and avoid unnecessary conversation.

Notice that descriptions and quantities on Figure 6-9 are listed in the same order on the Estimate Detail Form as they are on the Estimate Take-off Sheet. The first estimate category in Figure 6-9 is "Foundations." That shouldn't be a surprise. Foundations are the first heading on the Estimate Take-off Sheet, also.

The first cost in this section is foundation layout: setting batterboards and chalk-marking the excavation. The number 62 in the *Quantity* column comes from the *Total /Unit* column near the top of Figure 6-3. This 62 units consists of 14 corners and 12 footings, which are considered to be four corners each. The next column is the *Unit* column, meaning unit of measure. In this case, the unit is *Ea*, meaning *each*.

Estimate Detail Form

(Totals and descriptions carried forward from Estimate Take-off Sheet)

Company: SCC Contractors | **Estimate #** 06-1 | **Estimator:** Chris | **Date:** 5/19/93

Job: Sunny Hills Shopping Center Building #1 | **Estimate due:** 5/28/93 | **Checked by:** Kevin | **Date:** 5/21/93

Address: State Rt. 70 & Old Church Road

Job Description: Dave Phelps General Contractor Job # 93-05-12

Notes: See drawing (Figure 6-2). Foundations and curbs only. Slab-on-grade is an alternate bid item. Electric equip pad NOT included. Site preparation and control points by others, NOT included.

CSI Division/Account: 3.0/Concrete

Total base bid work, using a 2-man crew, yields 11 working days

Use 2-man crew: Chris at $24 per hour and Keith at $20 per hour = $22 per manhour average

Item or Description	Qty	Unit	Material Unit $	Material Ext $	Manhours MH/Unit	Manhours MH Ext	Labor MH $	Labor Ext $	Equipment Unit $	Equipment Ext $	Subcontract Unit $	Subcontract Ext $	Total Cost
Foundations													
Layout foundations, set batterboards and chalk mark excavations	62	Ea	2.25	$140	0.258	16	22.00	$352	--	--	--	--	$492
Excavation, use backhoe	42	CY	--	--	0.133	6	22.00	$132	1.00	$42	--	--	$174
Ready-mix concrete 2000 psi design mix 1½" aggregate placed directly from chute of truck	56	CY	47.00	$2,632	0.552	31	22.00	$682	0.95	$53	--	--	$3,367
Reinforcing steel bars ASTM A 615													
#4 & smaller Grade 40	1291	Lbs	0.30	$387	0.01	13	22.00	$286	--	--	--	--	$673
#5 & larger Grade 60	2882	Lbs	0.30	$864	0.01	29	22.00	$638	--	--	--	--	$1,502
Forms, Concrete placed directly against earth, no side forms or finishing required. Haunches and blockouts only													
Use double 2" x 6", base on 2 uses	2273	LF	0.18	$409	0.019	43	22.00	$946	--				$1,355
Strip and clean haunch forms	2273	LF	--	--	0.01	23	22.00	$506	--				$506
Backfill and minimum compaction (wheel roll)													
Around foundations, use backhoe	4	CY	--	--	Allow	1	22.00	$22	--	--	--	--	$22
Stockpile excess excavation on site	38	CY	--	--	0.023	1	22.00	$22	--	--	--	--	$22
Total foundations	56	CY	--	$4,432		163	22.00	$3,586		$95	--	--	$8,113
Curbs, straight, formed, 2000 psi ready-mix concrete placed directly from chute of ready-mix truck Use SCC standard assembly price for cast-in-place concrete curbs per linear foot													
6" wide x 12" high	22	LF	2.00	$44	0.109	2	22.00	$44	--	--	--	--	$88
6" wide x 6" high	28	LF	1.50	$42	0.109	3	22.00	$66	--	--	--	--	$108
Total straight curbs	50	LF		$86		5		$110		--		--	$196
Alternate bid item													
Use SCC price per SF for 4" slab-on-grade. Use 4-man crew. Chris at $24 per hour, Keith at $20 per hour and 2 laborers at $12 per hour each. Avg cost per manhour $17													
Alternate, Add 4" slab-on-grade	5000	SF	1.31	$6,550	0.020	100.00	17.00	$1,700	0.130	$650	--	--	$8,900

For scheduling purposes estimate slab with 4-man crew will require 3 working days

Costs shown DO NOT include overhead or profit

Estimate Detail Form 1 of 1

Figure 6-9 Estimate detail form

If you compare the Estimate Take-off Sheet to the Estimate Detail Form, you see that all the information from Figures 6-3 and 6-8 appears on Figure 6-9, and in the same order. Transfer everything first to the *Item or Description*, *Quantity* and *Unit* columns. Then go back and fill in the prices.

Material Cost Estimates

Chris based material prices for lumber, concrete and steel on supplier quotes. For example, a quick call to Mobile Transitcrete confirmed that 2000 psi mix delivered to the Sunny Hills site would cost $47 per cubic yard.

Notice that Chris estimated the cost of 2 x 6 lumber at only $0.18 per linear foot. That's only $180 per 1,000 board feet. Why so little? Chris knows that 2 x 6 lumber will cost him $360 per 1,000 board feet, or $0.36 per linear foot. But he also knows that he can use this form lumber several times. Because he allowed salvage value for this lumber, he'll only charge $0.18 per linear foot to the Sunny Hills job.

Manhour Estimates

Chris plans to use a two-man crew on this job: himself at $24 per hour and Keith at $20 per hour. That's an average of $22 per manhour, as noted in the heading in Figure 6-9.

Most labor costs in this estimate came from the cost records of previous jobs. But every job is unique. Even with good past cost records, don't price any estimate without considering what's different about *this* job.

For example, the first cost item on the estimate detail is for laying out the foundations. Chris knows that layout has averaged about one-tenth of a manhour per corner on previous jobs. For 62 corners, that would be only 6.2 manhours — or about 3 hours for a crew of two. In Chris' opinion, that was not enough time for this job. He felt an 8-hour day for a crew of two would be more realistic, so he wrote 16 in the *MH Ext* column. Working back to the *MH/Unit* column, the time per corner is 16 divided by 62, or 0.258 manhours per unit. The material cost for this line is based on previous jobs, and covers batterboards, stakes, and nails.

The column headed *MH Ext* is the manhours extension: the time per unit multiplied by the number of units. For example, Chris estimated that it would take 0.133 manhours to excavate each cubic yard with a backhoe. Forty-two cubic yards times 0.133 manhours is 5.59 manhours. Round up to the next full hour, in this case, 6 manhours.

Chris estimated that foundations would take a total of 163 manhours, and curbs would require 5 manhours. In the *MH Ext* column, see the estimate of 163 hours for Total foundations and 5 hours for Total straight curbs. That's 168 hours, or 21 man-days. A man-day is one man working one 8-hour day. Two men working 8 hours per day yields 16 manhours per day. Dividing 168 manhours by 16 gives us 10.5 working days. For scheduling purposes, Chris rounded that to 11 days for the crew of two.

When you estimate a job, use a wage rate based on the average for the crew. On this job, Chris will use a two-man crew. Although, for the backhoe operation, only one man can operate the backhoe at a time, the other man will still be on the job, perhaps with a shovel or pry bar, helping to get the job done. So normally, when it comes time to schedule this job, Chris will not assume he has another man available to do other work.

"Good manhour estimates are the key to good estimates."

Chris has learned that a cost estimate without a manhour estimate isn't very useful. He needs to know both how much the job will cost and how long it will take. Chris includes manhours in his estimates because his time estimate helps confirm total cost. If the estimated cost and the estimated manhours don't seem consistent with a previous, similar job, something is probably wrong.

Good manhour estimates are the key to good estimates. Most bidders have material costs that are within a few percent of being identical on most jobs. Labor estimates are different. There's room for legitimate disagreement on productivity estimates. In fact, there's no single "right" manhour estimate for most jobs. Every labor estimate has to be custom-crafted for the job, the crew, the supervision and the work conditions. That's why Chris shows a manhour estimate for every cost estimate in Figure 6-9.

Equipment unit costs usually come from SCC cost records and the accountant's forecast of the operating cost for SCC's backhoe-loader. They would normally be much higher than this. But Chris knows this job is a very small one for SCC, and he also knows the "going rate" per cubic yard for concrete. He decided to leave the equipment cost as is for now, and check it again before he finished the estimate.

Extending Costs and Manhours

The columns headed *Ext $* are the extended material, labor and equipment costs. (There's also an *Ext $* column for Subcontract work, but that doesn't apply to this job.) Extended costs are the cost per unit times the number of units. For example, Chris estimated that the material cost per corner for layout would be $2.25 each. (As we mentioned earlier, that covers the stakes, batterboards, and nails.) Multiply 62 corners by $2.25 for an extended cost of $140.

To extend labor costs, Chris used an average hourly rate of $22.00 per manhour, as you see in the note immediately above the *Item or Description* column.

Totaling the Columns

Once you've extended the manhours and costs, total the extension columns to find total costs and total manhours. Then add the extended costs in each row and write the sum in the *Total Cost* column at the far right. Add all the numbers in the *Total Cost* column. Finally, add the column totals across to verify the total in the *Total Cost* column.

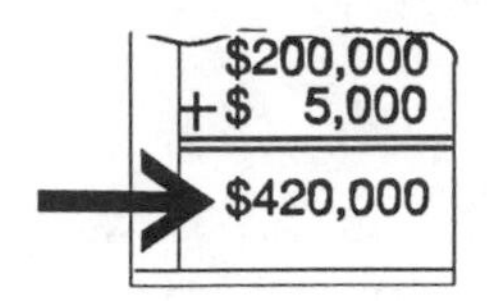

For example, under foundations, $4,432 plus $3,586 plus $95 is $8,113. Add the figures in the *Total Cost* column for foundations and you'll get the same number. This is called checking with a *crossfoot*. The column total and the row total should be the same. If the crossfoot doesn't add up correctly, something is wrong. In this case, it checked perfectly.

The Estimate Summary

The estimate summary is a convenient reference

The Estimate Summary Form (Figure 6-10) is where you collect and summarize costs from the Estimate Detail Form. In our example, the Detail form is only one page, but in an actual estimate, the Estimate Detail may run to several pages. The Estimate Summary is a convenient reference, a recap of the estimate by category. In this case, there were only three categories: foundations, curbs, and the alternate bid for the slab.

Write the same information at the top of the Estimate Summary as appears on the Take-off and the Estimate Detail. Be consistent when you identify the parts of your estimates. That makes it easier when it's necessary to follow the audit trail back to the source documents. It's easy to mis-file estimates, especially if you're working on more than one estimate at a time. It's a good idea to write the estimate number on the bottom of the forms as well.

Lines in the heading identify the scope of work and the scheduled completion time. To emphasize the procedure he followed, Chris made a note on the Estimate Summary: *Carried forward from Estimate Detail Sheet.*

The column at the left identifies the source of information in each estimate category. That's important in a larger estimate that may have dozens of estimate detail pages. The next column to the right shows the item description. Columns to the right show total material, labor and equipment costs and labor manhours. If there were any subcontract costs in this job, they would appear in the column headed *Subcontract Cost*. The column at the far right shows the total for each row.

Check your math

Once again, check the math by adding costs both vertically and horizontally. If vertical and horizontal totals don't match, look for a mistake somewhere.

Estimate Summary Form

(Carried forward from Estimate Detail Sheet)

Company: SCC Contractors
Estimator: Chris Riley
Date: 5/21/93

Job: Sunny Hills Shopping Center Building #1
Checked by: Kevin Riley
Date: 5/25/93

Address & phone number: State Rt. 70 & Old Church Road, 555-1234
Estimate # 06-1

Job Description: Dave Phelps General Contractor Job #93-05-12
Estimate due: 5/28/93

Scope of Work: Foundations and Curbs only. For 5,000 SF single story masonry block building. 4" thick slab-on-grade bid as an alternate.

CSI Division / Account: 3.0/Concrete Foundations & Straight Curbs

Scheduled completion, base bid: 2-man crew 11 working days after start of work
Scheduled completion, alternate bid: 4-man crew 3 working days after start of work

Estimate Detail Page & Account	Item Description	Material Cost	Labor Manhours	Labor Cost	Equip Cost	Sub-contract Cost	Total Cost
	Base Bid:						
Page 1 of 1 3.0 Concrete	Foundations (56 CY concrete = $144.88 per CY)	$4,432	163	$3,586	$95	--	$8,113
	Straight curbs (50 LF yields $3.92 per LF)	$86	5	$110	--	--	$196
	Subtotal direct costs, Base Bid	$4,518	168	$3,696	$95	--	$8,309
Markup	Overhead, profit and supervision at 30%	$1,355	50	$1,109	$29	--	$2,493
Contingency at 5%	Allow for job size, lack of detail on drawings	$294	11	$240	$6	--	$540
	Total base bid, all of the above	$6,167	229	$5,045	$130	--	$11,342
			Total base bid per plans & specs				$11,342

Estimate Detail Page & Account	Item Description	Material Cost	Labor Manhours	Labor Cost	Equip Cost	Sub-Contract Cost	Total Costs
	Alternate bid, add for floor slab						
Page 1 of 1 3.0 Concrete	(5,000 SF = $1.78 per SF)						
	Add for 4" thick slab-on-grade, 5,000 SF	$6,550	100	$1,700	$650	--	$8,900
Markup	Overhead, profit and supervision at 30%	$1,965	30	$510	$195	--	$2,670
Contingency at 5%	Allow for job size, lack of detail on drawings	$426	7	$111	$42	--	$579
	Total alternate bid, all of the above	$8,941	137	$2,321	$887	--	$12,149
	Schedule completion, 4-man crew, 3 working days	**Total alternate, for 4" slab-on-grade add**					$12,149

Bid alternate as: 4" concrete slab w/6x6 - 10/10 W.W.F. over 2" sand, unit price at $2.43 per SF in place.

Suggest that our bid include a price for all work, as described, on an uninterupted schedule at $23,000 lump sum

SCC Estimate #06-1 Estimate Summary Page 1 of 1

Figure 6-10 Estimate summary form

Overhead and Profit

Chris stopped when he finished work down to the "Base Bid" subtotal lines in the Estimate Summary. The rest was up to Kevin.

Follow up on questions

Kevin analyzed the estimate, studied the plans and specifications, and reviewed the manhour estimates. Next, he called Phelps Construction to ask about the storm drain. (That call was to follow up on his note on the sketch he made at the site visit.) He was told that off-site drainage would be finished before foundation work began.

Kevin told Phelps that SCC would bid the slab as an alternate and could shave a little off the price if they could do both jobs on an uninterrupted schedule. Phelps couldn't commit to that, but invited Kevin to add the slab as an alternate on his bid.

It's Kevin's job to decide how much to add to the estimate for overhead and profit. There were two things he didn't like about the Sunny Hills project. The first was the most obvious: it was smaller than most SCC jobs. If SCC was busy, he wouldn't have bothered to bid phase one of Sunny Hills. But SCC wasn't busy and Kevin wanted to land some extra work.

Second, he had doubts about the developer. The firm was from out of state and hadn't done any work in Alabama before. He didn't know anything about their payment habits or their reputation for dealing with contractors. Even though Kevin felt comfortable working with Phelps Construction, he was apprehensive about what the owner's representative might demand.

Kevin added 18 percent to the bid for overhead and another 12 percent for profit. Then, because the job was small, and because he didn't know much about the developer, Kevin added 5 percent for contingency. This markup is on the high side for SCC. But, although they can use the work during the upcoming slack time, they're not willing to bid the job at a loss. We'll tell you more about those considerations in Chapter 11.

Kevin wrote in the figures for overhead, profit and contingency and added the rows and columns. He realized this estimate of $148 per cubic yard for the concrete work ($8,309 divided by 56) was at the high end of their usual range for this work ($135 to $150), so he decided not to raise the figure for equipment charged to this job.

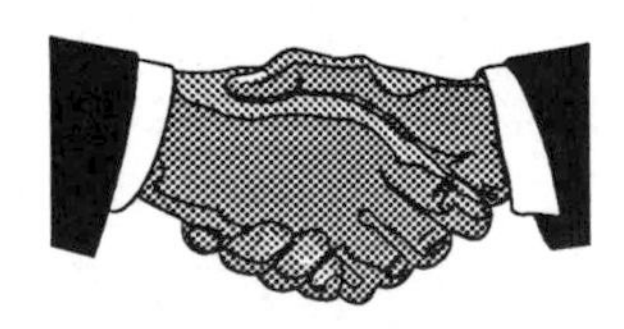

Kevin was sure the estimate was perfect. But he left it on his bookkeeper's desk for her to check one last time. Ruthie's sharp eye had found mistakes in "perfect" estimates before. This time she found none, so she typed a fresh copy of the estimate on SCC's standard proposal form and faxed it to Phelps Construction. Eventually Phelps won the contract for Sunny Hills, and awarded the concrete subcontract to SCC.

Had Kevin and Chris prepared this estimate on their computer, Ruthie would simply have added some formatting, then printed a copy for Phelps on the SCC company letterhead. I'll show you how that works in the next chapter.

Chapter 7
Pricing With a Computer

Refer back to Figures 6-8 and 6-9 in Chapter 6 and you'll see neatly typed cost estimates. That's a little deceptive. Chris actually wrote out these estimates in pencil. Then Ruthie used Microsoft Excel to type the forms you see in Chapter 6. If that seems like double work to you, Chris has an excellent answer. It's what works for SCC Construction. In Chris' words, "If it ain't broke, don't try to fix it," And he's probably right — at least for SCC Construction.

But that's not the whole story. SCC Construction is no stranger to computers. In fact, it was 1984 when the first computer arrived in SCC's office. In the spring of 1986 Kevin bought a suite of construction management programs from Orion Designs, an Atlanta software developer. Auto-Mate for Builders included modules for scheduling, payroll, accounting, estimating, job costing, purchase orders, receivables and payables. Ruthie uses the accounting module every month. But that's the only part of Auto-Mate currently in use. Kevin tried the estimating and scheduling programs and found fault with each. Eventually he went back to estimating with pencil and paper. The other programs really never got a fair trial in the SCC office. Kevin insists that Auto-Mate for Builders was an excellent value at $2,995. The fault was his, he admits. In Kevin's words, "I bought more software than I could learn."

Kevin got an early start with computers. But he's a little behind now. According to one recent survey of small and mid-size construction companies, 57% are using a computer for estimating. The most popular uses for a computer in construction offices are:

Word processing	76%
Bookkeeping or accounting	60%
Construction cost estimating	57%
Drafting, drawing or CAD	39%
Accounts receivable	33%

If, like Chris, you aren't preparing estimates with a computer, here's your chance. As you've probably noticed already, there's a compact computer disk called CD Estimator inside the back cover of this book.

Using CD Estimator

CD Estimator has an exceptionally large database – over 2,500 pages of 1996 labor, material and equipment costs. You can page through these costs one screen at a time or use the electronic index to search by keyword for exactly the information you need. When you find what you're looking for, split the screen in two so your estimate is on the bottom half of the screen and the database is on the top half. Then press a key to pull a cost estimate out of the database and into your estimate. Type the estimated quantity. The program extends prices and totals columns automatically. Add your overhead and profit as a percentage and then print the estimate.

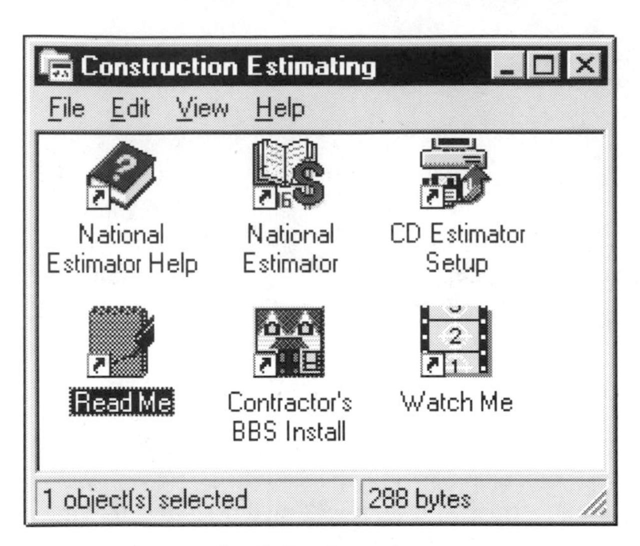

Construction Estimating program group

If you're a Windows user, you know most of the National Estimator program already. But even computer novices can be writing labor and material cost estimates in an hour or two. CD Estimator may be exactly the program you've been looking for. The only way to be sure is to give it a try. The first step is installing the CD Estimator disk.

The installation program creates a Construction Estimating program group and puts six icons in that group. Information on the next page explains how to use each of these icons. If you have trouble installing CD Estimator, call Craftsman tech support at 619-438-7828.

What's on the CD Estimator Disk

Here's a roadmap to what you'll find on CD Estimator:

National Estimator

National Estimator is the heart of the CD Estimator disk. It's a neat little construction cost estimating program for Windows™. You can use National Estimator all by itself. But it's the perfect way to use the six construction cost estimating databases on CD Estimator. When installation of CD Estimator is complete, click on the National Estimator icon to begin running the program. The next time you want to use National Estimator, click on Start, click on Programs, click on Construction Estimating and then on National Estimator. Instructions for using National Estimator begin on page 156.

Over 2,500 pages of 1996 labor and material costs from six costbooks are on the CD Estimator disk. When National Estimator is running, any of the six costbooks are just a mouse click away. Click on File, click on Open Costbook, then select the costbook you need. The file name for each of the six costbooks is listed below. You can open all six at once and use labor and material costs from all six in a single estimate. The six 1996 construction estimating costbooks are:

96CONST.CBK is the 1996 *National Construction Estimator*
96REP.CBK is the 1996 *National Repair & Remodeling Estimator*
96ELECT.CBK is the 1996 *National Electrical Estimator*
96PLUMB.CBK is the 1996 *National Plumbing & HVAC Estimator*
96PAINT.CBK is the 1996 *National Painting Cost Estimator*
96INSUR.CBK is the 1996 *National Renovation & Insurance Repair Estimator*

Watch Me

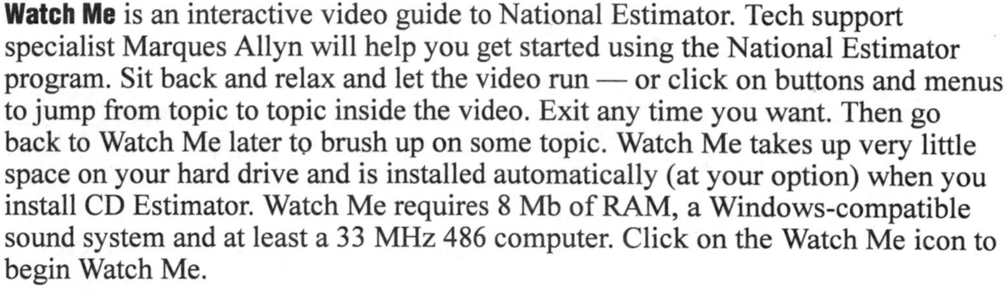

Watch Me is an interactive video guide to National Estimator. Tech support specialist Marques Allyn will help you get started using the National Estimator program. Sit back and relax and let the video run — or click on buttons and menus to jump from topic to topic inside the video. Exit any time you want. Then go back to Watch Me later to brush up on some topic. Watch Me takes up very little space on your hard drive and is installed automatically (at your option) when you install CD Estimator. Watch Me requires 8 Mb of RAM, a Windows-compatible sound system and at least a 33 MHz 486 computer. Click on the Watch Me icon to begin Watch Me.

Excalibur Client

Contractor's BBS

Excalibur is a communications program for Windows. Costs on your CD Estimator disk are revised quarterly and made available at no charge on Contractor's Bulletin Board. You pay only for the phone call. Use Excalibur to log on to Contractor's BBS and download both updated costs and new releases of National Estimator. To use Excalibur, you'll need a modem (9600 baud or faster). The Excalibur program isn't installed immediately when you install CD Estimator. Later, when you're ready to install Excalibur, double click on The Contractor's BBS Install icon. That begins installation of Excalibur and puts the Excalibur Client icon in your Construction Estimating program group. When you're ready to install Excalibur, turn to page 168.

Forty Estimating and Bidding Forms are included on CD Estimator. These forms are on the FORMS directory on CD Estimator. If you have a laser or ink jet printer, you'll be able to create top-quality customized forms in minutes. To use these forms, you'll need any of the popular word processing or spreadsheet programs. First, start your word processing or spreadsheet program. Then click on File and Open. Change the drive letter to your CD (usually D:) and the directory to FORMS. You'll see over forty forms created especially for each of the following programs:

- Microsoft Word forms have a file name extension of DOC
- Microsoft Excel forms have a file name extension of XLS.
- Lotus 1-2-3 forms have a file name extension of WK3
- Microsoft Works forms have a file name extension of WKS
- WordPerfect forms have a file name extension of WP

To open any form, double click on the form name. Make the changes you want. Then save the modified form to your hard drive (such as C:). For a description of all forms on the CD Estimator disk, open the file INDEX.TXT on the FORMS directory.

CD Estimator Setup

CD Estimator Setup gets you back into the installation program if you decide to install more CD Estimator components later. For example, we recommend that you leave the six costbooks on your CD Estimator disk, at least in the beginning. Most users find that National Estimator runs fast enough even if the six costbooks aren't installed on their hard drive. Leaving the six costbooks off your hard drive saves about 12 Mb of valuable space. If there's space to burn on your hard drive, go ahead and install all six costbooks. National Estimator may run a little faster. Either way, the CD Estimator disk has to be in the CD-ROM drive when CD Estimator is running.

National Estimator Help

National Estimator Help has everything you could every want to know about National Estimator. These help files are available any time National Estimator is running (click on the question mark) or from the Construction Estimating program group (click on National Estimator Help). To print a 27-page user's guide to National Estimator, click on the question mark. Then click on Print All Topics. At Guide to National Estimator, click on File. Finally, click on Print Topic.

Read Me

Read Me has notes and suggestions that might be of interest. Most users can safely ignore Read Me. If you're curious, click on the Read Me icon.

Now that you know what's on the CD Estimator disk, let's get started. The first step is installation.

Installing CD Estimator with SETUP.EXE

Put the CD Estimator disk in your CD drive (such as D:). Start Windows.

In Windows 3.1 or 3.11
go to the Program Manager.

1. Click on File.
2. Click on Run . . .
3. Type D:\SETUP
4. Press [Enter ↵].

In Windows 95,

1. Click on Start.
2. Click on Run . . .
3. Type D:\SETUP
4. Press [Enter ↵].

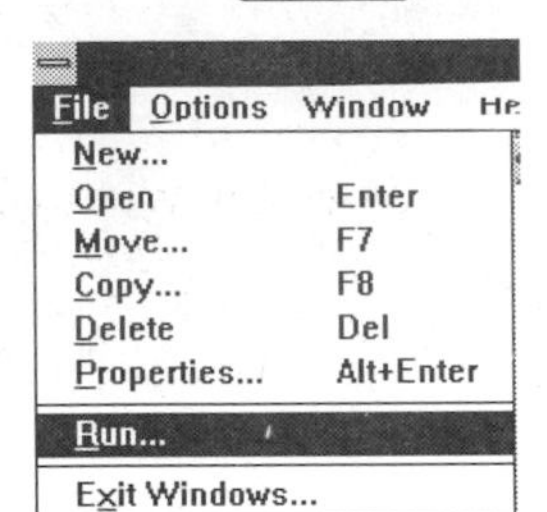

Click on File. Click on Run

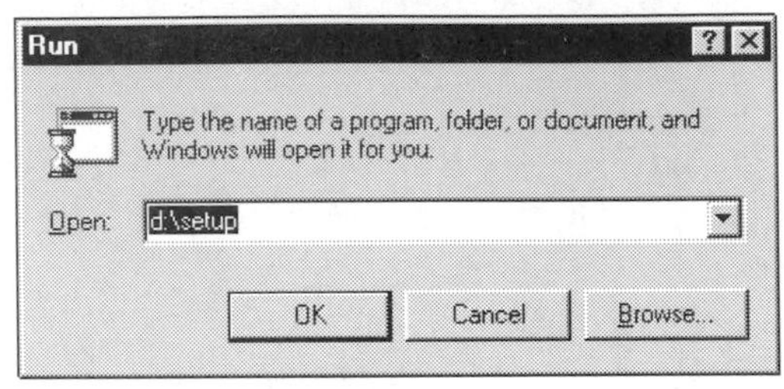

Type D:\SETUP in Windows 95.

Selecting the Files and Directory for Installation

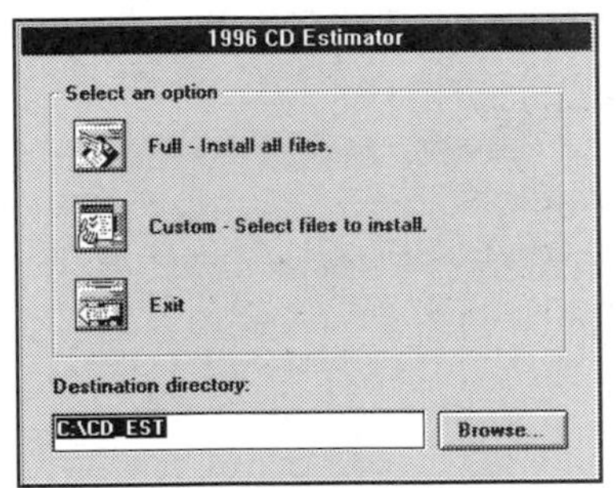

Clicking on "Custom" is recommended.

Decide on either Full or Custom installation. To make your selection, just click on the button of your choice:

- **Full - Install all files** is the best choice if there's plenty of space available on your hard drive. Full installation copies all six costbooks to the hard drive of your computer. That's about 15 Mb.
- **Custom - Select files to install** is the recommended choice. Most users find that National Estimator runs fast enough with costbooks on the CD Estimator disk. If you have a slow CD-ROM drive (single speed), performance will probably be better if costbooks are copied to the hard drive. In any case, CD Estimator has to be in the CD-ROM drive when you're using CD Estimator.
- **Destination directory C:\CD_EST** is recommended. To install to some other drive or directory, click in the file name box and type the drive and directory of your choice.

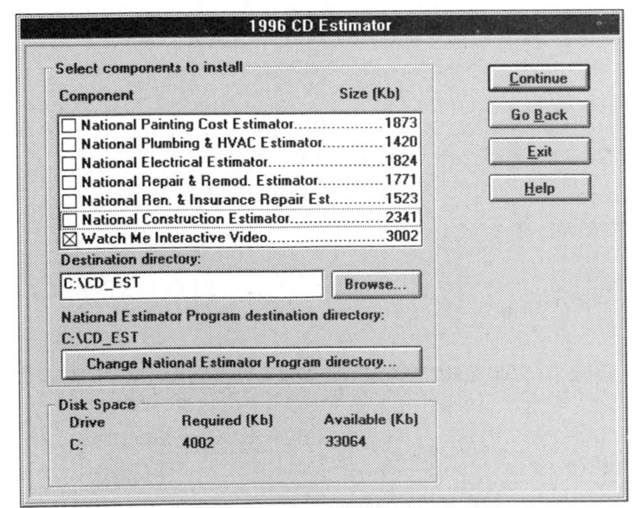

Recommended: Click on all six costbooks so none are copied to your hard drive.

If you selected Custom installation, the next step is identifying files to be installed. An X in the box at the left of a costbook name indicates the costbook will be copied to your hard drive. To remove the X, just click on the costbook name. We recommend that you click on all six costbooks so none are copied to your hard drive. See the illustration at the left.

If performance is unsatisfactory with costbooks on the CD, you can always restart CD Estimator Setup and install the costbooks you use most. Just click on CD Estimator Setup in your Construction Estimating program group. Then leave checked the costbooks you want to install and click on Continue.

Leave checked Watch Me Interactive Video if you plan to use Watch Me. This interactive video requires at least 8 Mb of RAM and a Windows-compatible sound system. Once you've learned National Estimator, you're done with Watch Me. For instructions on removing Watch Me files from your hard disk, click on Read Me in the Construction Estimating program group.

When finished selecting files, click on Continue to begin installation.

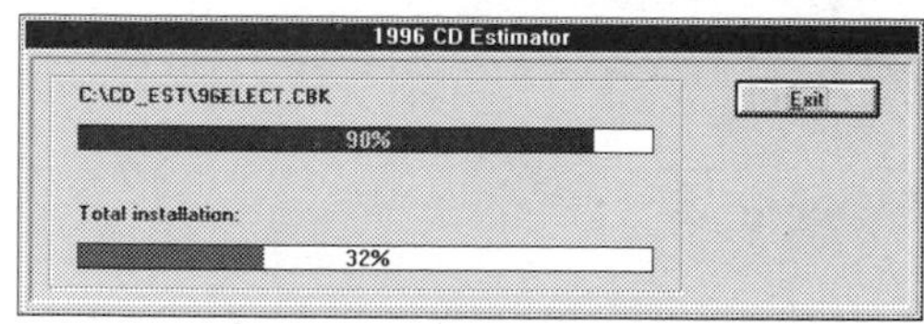

Installation takes 2 or 3 minutes.

Copying files to the hard drive will take two or three minutes on most computers. When files have been copied, you'll see six icons in the Construction Estimating program group.

The last step is installing Video for Windows (if you're a Windows 3.1 user).

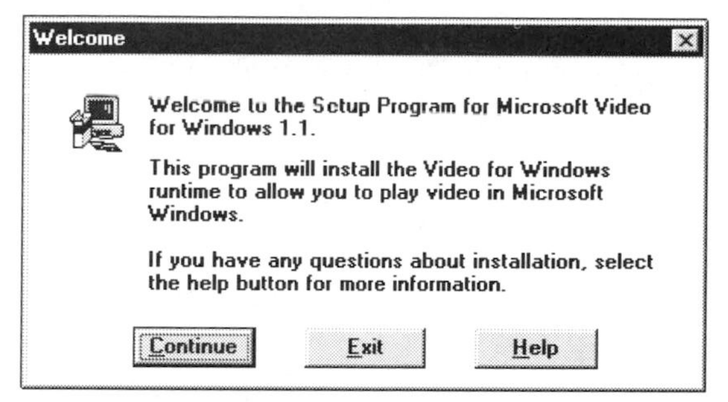

Windows 3.1 - Click on Continue.
Windows 95 - Click on Exit.

- If you're using Windows 3.1 and elected to install Watch Me, Microsoft Video for Windows will be installed next. Click on Continue. When Video for Windows has been installed, click on Restart Now so the changes will be in effect. When you see the Construction Estimating program group, you're ready to begin estimating.
- If you're using Windows 95, don't install Video for Windows. It isn't needed. Click on Exit to terminate installation. When you see the Construction Estimating program group, you're ready to begin estimating.

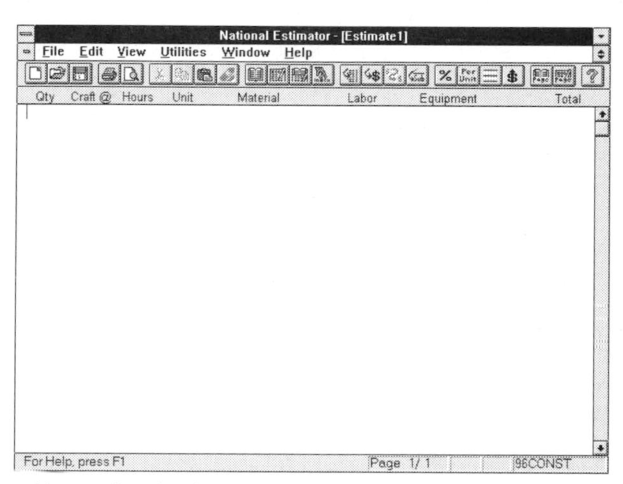

Your estimating form ready to begin the first estimate.

Using National Estimator

National Estimator Icon

Click on the National Estimator icon to begin using the program. In a few seconds your electronic estimating form will be ready to begin the first estimate.

On the title bar at the top of the screen you see the program name, National Estimator, and [Estimate1], showing that Estimate1 is on the screen. Let's take a closer look at other information at the top of your screen.

The Menu Bar

Below the title bar you see the menu bar: Every option in National Estimator is available on the menu bar. Click with your left mouse button on any item on the menu to open a list of available commands.

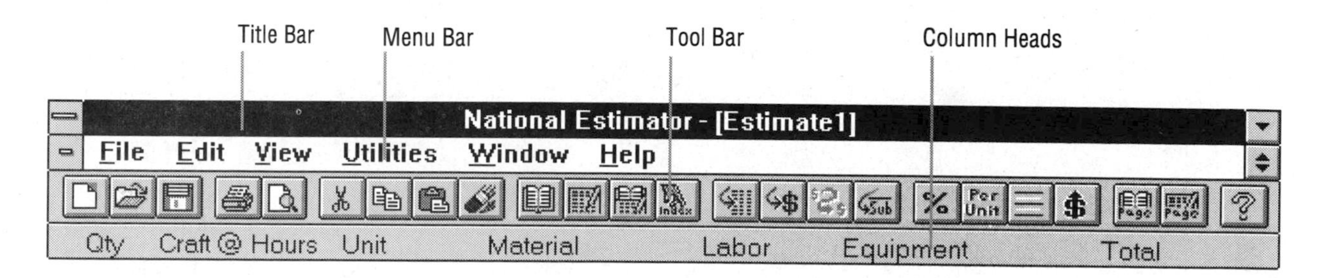

Buttons on the Tool Bar

Below the menu bar you see 24 buttons that make up the tool bar. The options you use most in National Estimator are only a mouse click away on the tool bar.

Column Headings

Below the tool bar you'll see column headings for your estimate form:

- **Qty** for quantity
- **Craft@Hours** for craft (the crew doing the work) and manhours (to complete the task)
- **Unit** for unit of measure, such as per linear foot or per square foot
- **Material** for material cost
- **Labor** for labor cost
- **Equipment** for equipment cost
- **Total** for the total of all cost columns

The Status Bar

The bottom line on your screen is the status bar. Here you'll find helpful information about the choices available. Notice "Page 1/ 1" near the right end of the status line. That's a clue that you're looking at page 1 of a one-page estimate.

Check the status bar occasionally for helpful tips and explanations of what you see on screen.

Beginning an Estimate

Let's start by putting a heading on this estimate.

The Blinking Cursor (insert point)

Mouse Pointer

Mouse Pointer

1. Press [Enter] once to space down one line.
2. Press [Tab] four times (or hold the space bar down) to move the cursor (the insert point) near the middle of the line.
3. Type "Estimate One" and press [Enter]. That's the title of this estimate, "Estimate One."
4. Press [Enter] again to move the cursor down a line. That opens up a little space below the title.

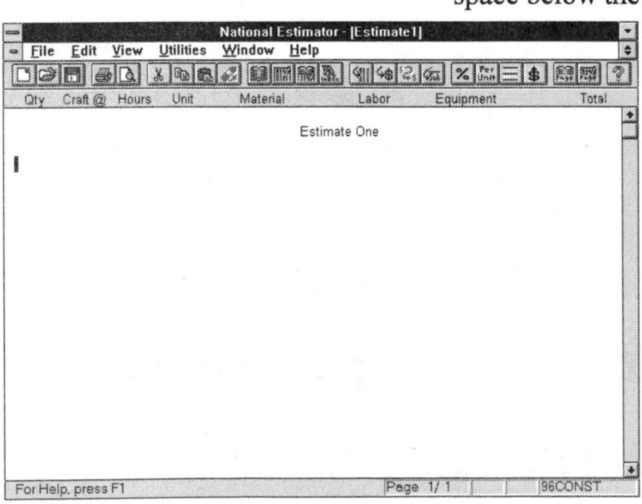

Begin by putting a title on your estimate, such as "Estimate One."

The Costbooks

Let's leave your estimating form for a moment to take a look at estimates stored in the costbooks. To switch to your default costbook, either:

- Click on the button, -Or-
- Click on View on the menu bar. Then click on Costbook Window, -Or-
- Tap the [Alt] key, tap the letter V (for View) and tap the letter C (for Costbook Window), -Or-
- Press [Esc]. Press the [↓] key to highlight Costbook Window. Then press [Enter].

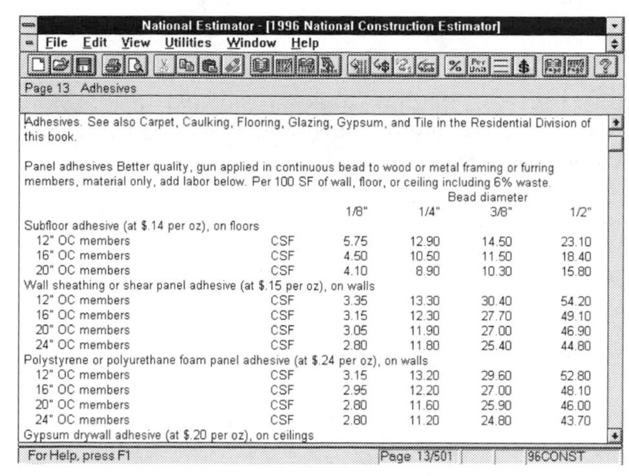

Adhesives. See also Carpet, Caulking, Flooring, Glazing, Gypsum, and Tile in the Residential Division of this book.

Panel adhesives Better quality, gun applied in continuous bead to wood or metal framing or furring members, material only, add labor below. Per 100 SF of wall, floor, or ceiling including 6% waste.

		Bead diameter			
		1/8"	1/4"	3/8"	1/2"
Subfloor adhesive (at $.14 per oz), on floors					
12" OC members	CSF	5.75	12.90	14.50	23.10
16" OC members	CSF	4.50	10.50	11.50	18.40
20" OC members	CSF	4.10	8.90	10.30	15.80
Wall sheathing or shear panel adhesive (at $.15 per oz), on walls					
12" OC members	CSF	3.35	13.30	30.40	54.20
16" OC members	CSF	3.15	12.30	27.70	49.10
20" OC members	CSF	3.05	11.90	27.00	46.90
24" OC members	CSF	2.80	11.80	25.40	44.80
Polystyrene or polyurethane foam panel adhesive (at $.24 per oz), on walls					
12" OC members	CSF	3.15	13.20	29.60	52.80
16" OC members	CSF	2.95	12.20	27.00	48.10
20" OC members	CSF	2.80	11.60	25.90	46.00
24" OC members	CSF	2.80	11.20	24.80	43.70
Gypsum drywall adhesive (at $.20 per oz), on ceilings					

The costbook window has the entire *National Construction Estimator.*

All six costbooks, are available in the Costbook Window. *National Construction Estimator* is the default costbook and opens automatically. If you have a color monitor, you'll notice right away that cost lines are blue and text (descriptions) are black. Notice also the words *Page 13 Adhesives* at the left side of the screen just below the tool bar. That's your clue that the adhesives section of page 13 is on the screen. To turn to the next page, either:

- Press PgDn (with Num Lock off), -Or-
- Click on the lower half of the scroll bar at the right edge of the screen.

To move down one line at a time, either:

- Press the ↓ key (with Num Lock off), -Or-
- Click on the arrow on the down scroll bar at the lower right corner of the screen.

Press PgDn about 1,800 times and you'll page through the entire *National Construction Estimator*. Obviously, there's a better way. To turn quickly to any page, either:

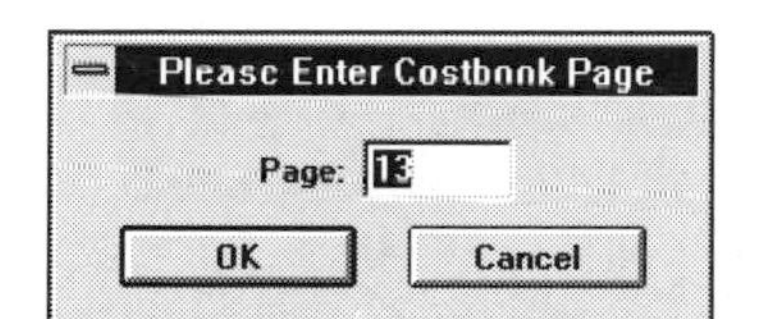

Type the page number you want to see.

- Click on the button at the right end of the tool bar, -Or-
- Click on View on the menu bar. Then click on Turn to Costbook Page, -Or-
- Tap the Alt key, tap the letter V (for View) and tap the letter T (for Turn to Costbook page), -Or-
- Press Esc. Press the ↓ key to highlight Turn to Costbook Page. Then press Enter←.

Type the number of the page you want to see and press Enter←. National Estimator will turn to the top of the page you requested.

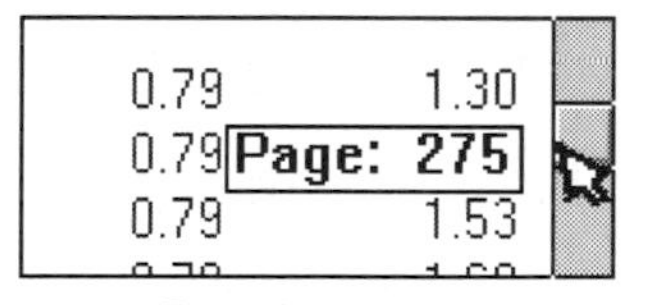

Drag the square to see any page.

An Even Better Way

Find the small square in the slide bar at the right side of the Costbook Window. Click and hold on that square while rolling the mouse up or down. Keep dragging the square until you see the page you want in the Page number box. Release the mouse button to turn to the top of that page.

A Still Better Way: Keyword Search

To find any cost estimate in seconds, search by keyword in the index. To go to the index, either:

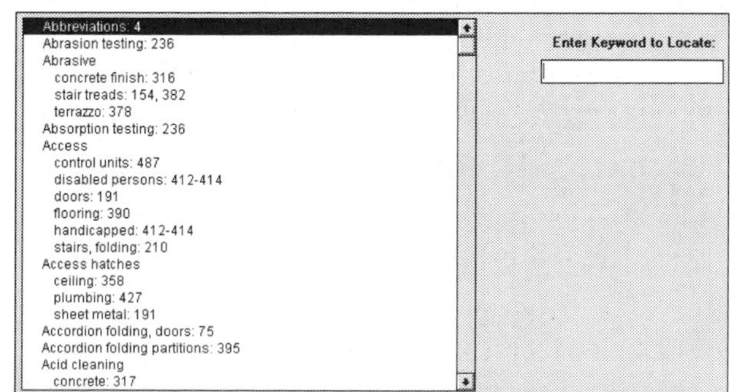

Use the electronic index to find cost estimates for any item.

- Click on the button near the center of the tool bar, -Or-
- Press Esc. Press Enter←, -Or-
- Tap the Alt key, tap the letter V (for View) and press Enter←, -Or-
- Click on View on the menu bar. Then press Enter←.

Notice that the cursor is blinking in the Enter Keyword to Locate box at the right of the screen. Obviously, the index is ready to begin a search.

Type the keyword to locate.

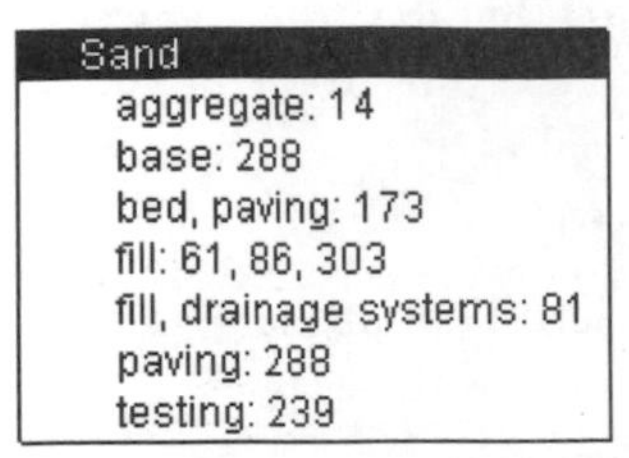

The index jumps to Sand.

Your First Estimate

Suppose we're estimating the cost of 4" sand fill under a 10′ x 10′ (100 square foot) concrete slab. Let's put the index to work with a search for sand fill. In the box under Enter Keyword to Locate, type *sand*. The index jumps to the heading *Sand*.

The fourth item under Sand is *fill: 61, 86, 303*. Either:

- Click once on that line and press [Enter←], -Or-
- Double click on that line, -Or-
- Press [Tab] and the [↓] key to move the highlight to *fill: 61, 86, 303*. Then press [Enter←].

If costs appear on several pages, click on the page you prefer.

To select the page you want to see (page 61 in this case), either:

- Click on number 61, -Or-
- Press [Tab] to highlight 61. Then press [Enter←].

National Estimator turns to the top of page 61. Notice that estimates at the top of this page are for slabs. "Slabs, walks and driveways" are at the top of the screen, as shown at the left below. Press the [↓] key (or click on the down scroll arrow) until "Slab Base" moves to the top half of your screen.

Splitting the Screen

Most of the time you'll want to see what's in both the costbook and your estimate. To split the screen into two halves, either:

- Click on the [button] button near the center of the tool bar, -Or-
- Press [Esc] and the [↓] key to move the selection bar to Split Window, -Or-
- Tap the [Alt] key, tap the letter V (for View), tap the letter S (for Split Window), -Or-
- Click on View on the menu bar. Then click on Split Window and your screen should look like the example at the right below.

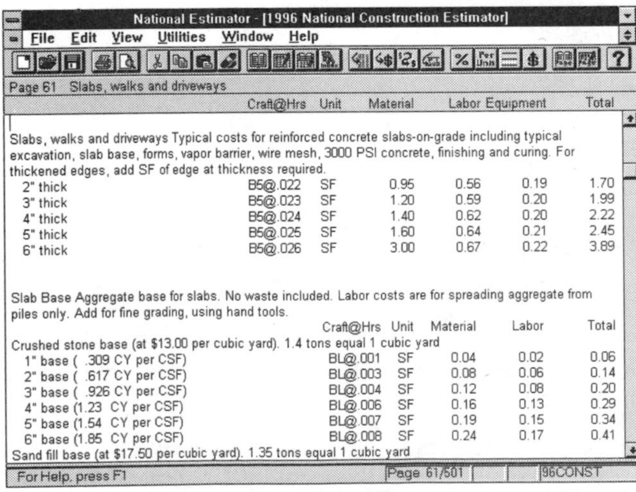

At the top of page 61, press the [↓] key to see sand fill.

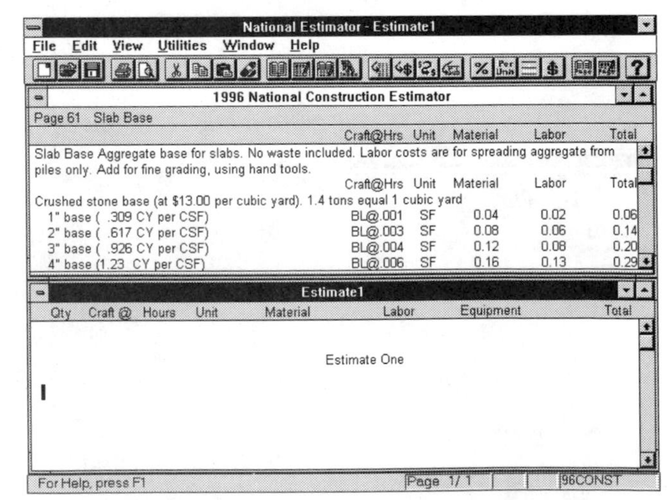

The split window: Costbook above and estimate below.

Notice that eight lines of the costbook are at the top of the screen and your estimate is at the bottom. You should recognize "Estimate One." It's your title for this estimate. Column headings are at the top of the costbook and across the middle of the screen (for your estimate).

To Switch from Window to Window

- Click in the window of your choice, -Or-
- Hold the Ctrl key down and press Tab.

Notice that a window title bar turns dark when that window is selected. The selected window is where keystrokes appear as you type. The estimate in the selected window is called the *current estimate*. If a costbook is in the selected window, it's called the *current costbook*.

Copying Descriptions to Your Estimate

We'll begin this estimate by copying the line that begins "Slab Base" to your estimating form. Click on the button on the tool bar to be sure you're in the split window. Click in the upper half of the split window. Then either:

- Move the cursor to the line "Slab Base" with the ↑ or ↓ key, -Or-
- Click with the left mouse button on the line "Slab Base."

There are four ways to copy the "Slab Base" line to the end of your estimate. Either:

- Press the F8 key, -Or-
- Click on the button on the tool bar. Then click on the button to paste, -Or-
- Hold the Ctrl key down and press C to copy. Then hold the Ctrl key down and press V to paste, -Or-
- Click on Edit on the menu bar. Then click on Copy Current Paragraph. Click on Edit again. Then click on Paste Into Current Estimate.

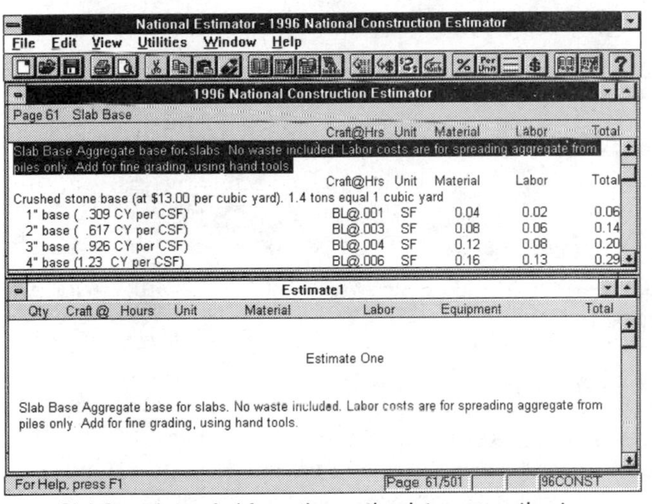

Slab base is copied from the costbook to your estimate.

Notice that the "Slab Base" line has been copied to your estimate.

Let's copy another line. Move the cursor down ten lines with the ↓ key so it's on the line that begins "Sand fill base. . ." Copy that line to the end of your estimate using any of the four copy methods just explained.

Inserting a Line in the Estimate

So far we've copied lines to the end of your estimate. You can insert a line anywhere in an estimate. Try inserting a line in the middle of the estimate. Here's how:

1. Begin by clicking again on the line that starts "Sand fill base . . ."
2. Click on the button.
3. Click in the estimate where you want the line inserted. (Any blank line will be OK.)
4. Click on the button.

Undo the Duplicate Paste

You've pasted the line "Sand fill base . . ." into your estimate twice. That's a mistake you can correct in the wink of an eye. Click in the lower half of the Split Window. Then either:

- Click on the button on the tool bar, -Or-
- Hold the Ctrl key down and type the letter Z, -Or-
- Click on Edit on the menu bar. Then click on Undo.

If the undo command or button is grayed out, it means the command isn't available any longer. You can undo many errors if you catch the mistake right away.

Copying Part of a Line

	BL@.007	SF
	BL@.008	SF
1.35 tons equal 1 cubic yard		
	BL@.001	SF
	BL@.002	SF

Select the words to copy.

You don't have to copy an entire line from the costbook to your estimate. You can copy a few words, an entire line or even several lines. Try copying just the words *1.35 tons equal 1 cubic yard* from the line that begins "Sand fill base." Begin by clicking just in front of the first character to select, the number 1. Then select the remaining words you want to copy. Either:

- Hold the left mouse button down while rolling your mouse to the right, -Or-
- Hold the Shift key down while pressing the → key.

The background behind the words turns dark as words are selected. When finished selecting:

1. Copy the selected words to the clipboard by clicking on the button on the tool bar.
2. Paste to the end of your estimate by clicking on the button on the tool bar. The words are pasted to your estimate.
3. Undo what you've just done. Click on the lower half of the Split Window. Then click on the button.

Copying Costs to Your Estimate

Next, we'll estimate the cost of 100 square feet of sand fill base. Click on the button on the tool bar to be sure you're in the split window. Click anywhere in the costbook (the top half of your screen). Then press the ↓ key until you see the cost estimate for 4" sand fill.

4" fill (1.23 CY per CSF)	BL@.004	SF	0.22	0.08	0.30

On a color monitor, this cost estimate line will be in blue to distinguish it from text lines which are black. Cost estimate lines are also in blue on your estimating form.

Enter Cost Information

Quantity:	1.00	OK
Unit of Measure:	SF	Cancel
Craft Code:	BL 1 building laborer	
Hourly Wage:	21.05	

	Unit Costs	Extended Costs
Hours:	0.004	0.00
Material Cost:	0.22	0.22
Labor Cost:	0.08	0.08
Equipment Cost:	0.00	0.00
Total Cost:	0.30	0.30

Use the Enter Cost Information dialog box to copy or change costs.

To copy this line to your estimate:

1. Click on the line.
2. Click on the button.
3. Click on the button to open the Enter Cost Information dialog box.

Notice that the blinking cursor is in the Quantity box:

1. Type a quantity of 100 because the slab base is 100 square feet (10′ x 10′).
2. Press Tab and check the estimate for accuracy.
3. Notice that the column headed Unit Costs shows costs per unit, per SF (square foot) in this case.

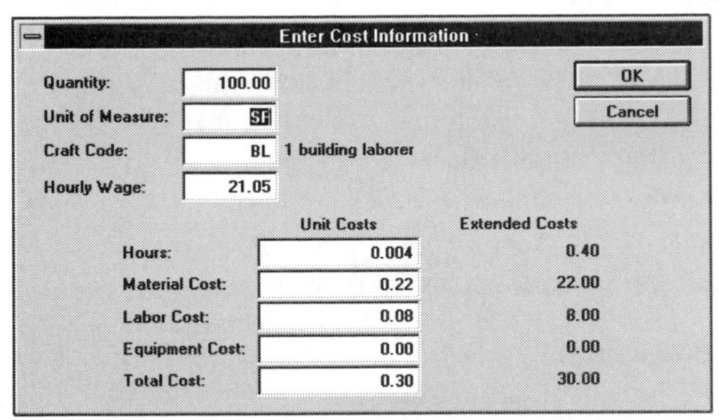

Costs for the 100 SF job (extended costs) are on the right.

4. The column headed Extended Costs shows costs for the entire slab base, 100 square feet.
5. Press [Enter↵] or click on OK to copy these figures to the end of your estimate.

The new line at the bottom of your estimate shows:

100.00 is the quantity of sand in square feet
BL is the recommended crew, a building laborer
@.4000 shows the manhours required for the work
SF is the unit of measure, square feet in this case
22.00 is the material cost (the sand fill)
8.00 is the labor cost for the job
0.00 shows there is no equipment cost
30.00 is the total of material, labor and equipment columns

4" fill (1.23 CY per CSF)						
100.00	BL@.4000	SF	22.00	8.00	0.00	30.00

Extended costs for sand fill base as they appear on your estimate form.

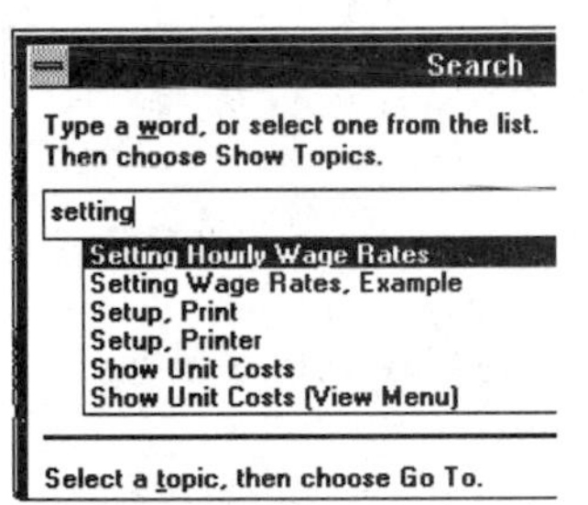

Search for information on setting wage rates.

Your Own Wage Rates

The labor cost in the example above is based on a building laborer working at an hourly cost of $21.05 per hour. (See pages 5 to 7 in the *National Construction Estimator* for crew rates used in the costbook.) Suppose $21.05 per hour isn't right for your estimate. What then? No problem! It's easy to use your own wage rate for any crew or even make up your own crew codes. To get more information on setting wage rates, press [F1]. At National Estimator Help Contents, click on the Search button. In the Search dialog box, type "setting" and press Enter twice. To return to your estimate, click on File on the National Estimator Help menu bar. Then click on Exit.

Making Changes in Your Estimate

Once you've copied a few lines to your estimating form, you'll probably discover that changes are needed. National Estimator makes it easy to change any line on the estimating form. First, click on the button on the tool bar to maximize the estimate window.

Changing Cost Estimates

With Num Lock off, use the [↑] or [↓] key to move the cursor to the line you want to change (or click on that line). In this case, move to the line that begins with a quantity of 100. To open the Enter Cost Information Dialog box, either:

- Press [Enter↵], -Or-
- Click on the button on the tool bar, -Or-
- Click on Edit on the menu bar. Click on Change a Cost Line.

	Unit Costs
Hours:	0.004
Material Cost:	0.18
Labor Cost:	0.08
Equipment Cost:	0.00
Total Cost:	0.26

Change the material cost to 18 cents.

To make a change, either

- Click on what you want to change, -Or-
- Press [Tab] until the cursor advances to what you want to change.

Then type the correct figure. In this case, change the material cost to .18 (18 cents).

Press [Tab] and check the Extended Costs column. If it looks OK, press [Enter↵] and the change is made on your estimating form.

Changing Text (Descriptions)

Click on the button on the tool bar to be sure you're in the estimate. With Num Lock off, use the ↑ or ↓ key or click the mouse button to put the cursor where you want to make a change. In this case, we're going to make a change on the line that begins "Sand fill base (at $17.50 per cubic yard) . . ."

To make a change, click where the change is needed. Then either:

- Press the Del or ←Bksp key to erase what needs deleting, -Or-
- Select what needs deleting and click on the button on the tool bar.
- Type what needs to be added.

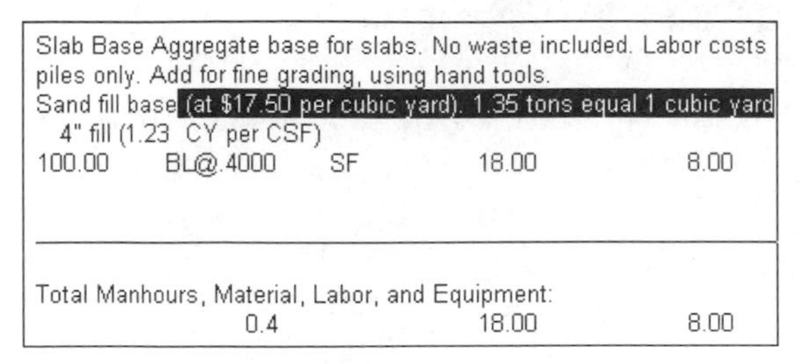

Slab Base Aggregate base for slabs. No waste included. Labor costs
piles only. Add for fine grading, using hand tools.
Sand fill base (at $17.50 per cubic yard). 1.35 tons equal 1 cubic yard
4" fill (1.23 CY per CSF)
100.00 BL@.4000 SF 18.00 8.00

Total Manhours, Material, Labor, and Equipment:
0.4 18.00 8.00

To select, click and hold the mouse button while dragging the mouse.

In this case, click after the word "base." Then hold the left mouse button down and drag the mouse to the right until you've put a dark background behind the words: "(at $17.50 per cubic yard). 1.35 tons equal 1 cubic yard." The dark background shows that these words are selected and ready for editing.

Press the Del key (or click on the button on the tool bar), and the selection is cut from the estimate.

Adding Text (Descriptions)

Some of your estimates will require descriptions (text) and costs that can't be found in the *National Construction Estimator*. What then? With National Estimator it's easy to add descriptions and costs of your choice anywhere in the estimate. For practice, let's add an estimate for four reinforced corners to Estimate One.

Click on the button to be sure the estimate window is maximized. We can add lines anywhere on the estimate. But in this case, let's make the addition at the end. Press the ↓ key to move the cursor down until it's just above the horizontal line that separates estimate detail lines from estimate totals. To open a blank line, either:

Slab Base Aggregate base for slabs. No
piles only. Add for fine grading, using ha
Sand fill base
4" fill (1.23 CY per CSF)
100.00 BL@.4000 SF
Reinforced corners

Total Manhours, Material, Labor, and Eq
0.4

Adding "Reinforced corners."

- Press Enter↵, -Or-
- Click on the button on the tool bar, -Or-
- Click on Edit on the menu bar. Then click on Insert a Text Line.

Type "Reinforced corners" and press Enter↵.

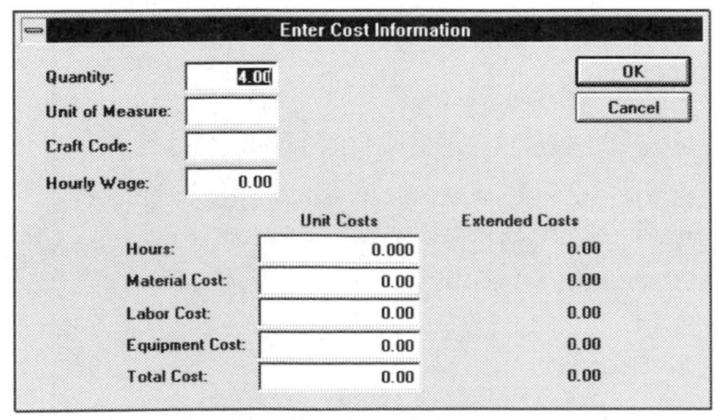

Enter Cost Information

Quantity: 4.00
Unit of Measure:
Craft Code:
Hourly Wage: 0.00

OK
Cancel

	Unit Costs	Extended Costs
Hours:	0.000	0.00
Material Cost:	0.00	0.00
Labor Cost:	0.00	0.00
Equipment Cost:	0.00	0.00
Total Cost:	0.00	0.00

Adding a cost line with the Enter Cost Information dialog box.

Adding a Cost Estimate Line

Now let's add a cost for "reinforced corners" to your estimate. Begin by opening the Enter Cost Information dialog box. Either:

- Click on button on the tool bar, -Or-
- Click on Edit on the menu bar. Then click on Insert a Cost Line.

1. The cursor is in the Quantity box. Type the number of units (4 in this case) and press Tab.

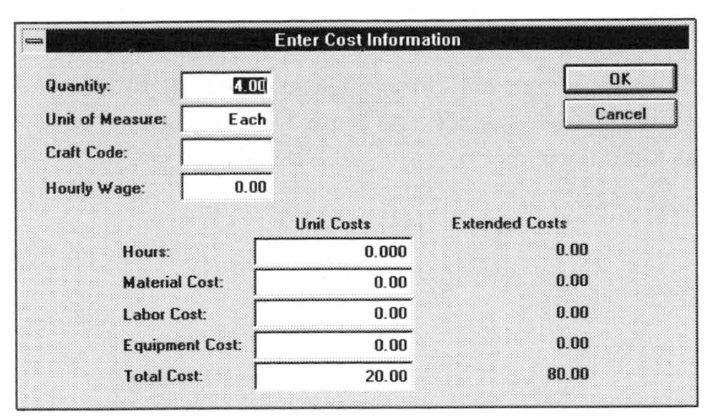

Unit and extended costs for four reinforced corners.

2. The cursor moves to the next box, Unit of Measure.
3. In the Unit of Measure box, type *Each* and press Tab.
4. Press Tab twice to leave the Craft Code blank and Hourly Wage at zero.
5. Since these corners will be installed by a subcontractor, there's no material, labor or equipment cost. So press Tab four times to skip over the Hours, Material Cost, Labor Cost and Equipment Cost boxes.
6. In the Total Cost box, type 20.00. That's the cost per corner quoted by your sub.
7. Press Tab once more to advance to OK.
8. Press Enter↵ and the cost of four reinforced corners is written to your estimate.

Note: The sum of material, labor, and equipment costs appears automatically in the Total Cost box. If there's no cost entered in the Material Cost, Labor Cost or Equipment Cost boxes (such as for a subcontracted item), you can enter any figure in the Total Cost box.

Adding Lines to the Costbook

Add lines or make changes in the costbook the same way you add lines or make changes in an estimate. The additions and changes you make become part of the user costbook attached to the *National Construction Estimator*. You may want to create several user costbooks, one for each type of work you handle. For more information on user costbooks, press F1. Click on Search. Type "user" and press Enter↵ twice.

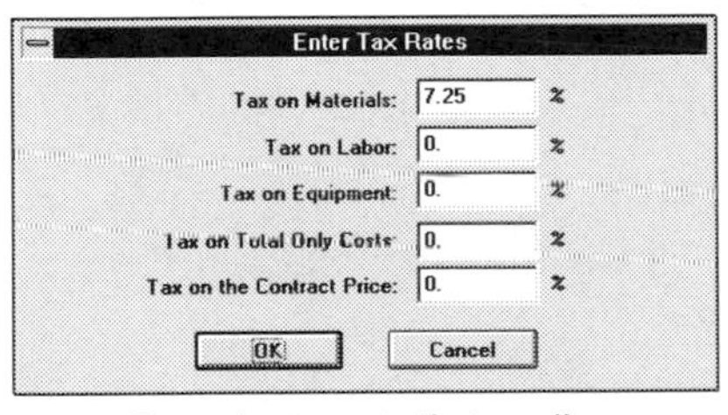

Type the tax rate that applies.

Adding Tax

To include sales tax in your estimate:

1. Click on Edit.
2. Click on Tax rates.
3. Type the tax rate in the appropriate box.
4. Press Tab to advance to the next box.
5. Press Enter↵ or click on OK when done.

In this case, the tax rate is 7.25% on materials only. Tax will appear as the last line of the estimate.

Adding Overhead and Profit

Add for Overhead & Profit

Overhead 15. %

Profit 10. %

OK Cancel

Adding overhead & profit.

Set markup percentages in the Add for Overhead & Profit dialog box. To open the box, either:

- Click on the $ button on the tool bar, -Or-
- Click on Edit on the menu bar. Then click on Markup.

Type the percentage you want to add for overhead. Press Tab. Type the percentage you want to add for profit. Press Enter↵ when done.

1. For this estimate, type 15 on the Overhead line.
2. Press Tab to advance to Profit.
3. Type 10 on the Profit line.
4. Press Enter↵.

Markup percentages can be changed at any time. Just reopen the Add for Overhead & Profit Dialog box and type the correct figure.

File Name: Estimate 1 — Construction Estimate — Page 1

Qty	Craft@Hours	Unit	Material	Labor	Equipment	Total
			Estimate One			
Slab Base Aggregate base for slabs. No waste included. Labor costs are for spreading aggregate from piles only. Add for fine grading, using hand tools.						
Sand fill base						
4" fill (1.23 CY per CSF)						
100.00	BL@.4000	SF	18.00	8.00	0.00	26.00
Reinforced corners						
4.00	--@ .0000	--	0.00	0.00	0.00	80.00
Total Manhours, Material, Labor, and Equipment:	0.4		18.00	8.00	0.00	26.00
Total Only (Subcontract) Costs:						80.00
			Subtotal:			106.00
			15.00% Overhead:			15.90
			10.00% Profit:			12.19
			Estimate Total:			134.09
			Tax on Materials:			1.30
			Grand Total:			135.39

A preview of Estimate 1

Preview Your Estimate

You can display an estimate on screen just the way it will look when printed on paper. To preview your estimate, either:

- Click on the button on the tool bar, -Or-
- Click on File on the menu bar. Then click on Print Preview.

In print preview:

- Click on Next Page or Prev. Page to turn pages.
- Click on Two Page to see two estimate pages side by side.
- Click on Zoom In to get a closer look.
- Click on Close when you've seen enough.

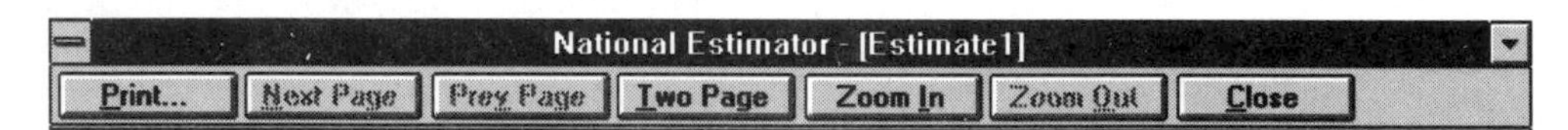

Use buttons on Print Preview to see your estimate as it will look when printed.

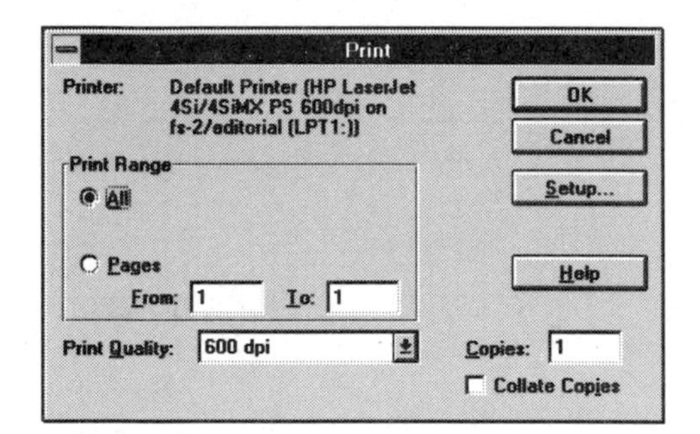

Options available depend on the printer you're using.

Printing Your Estimate

When you're ready to print the estimate, either:

- Click on the button on the tool bar, -Or-
- Click on File on the menu bar. Then click on Print, -Or-
- Hold the [Ctrl] key down and type the letter P.

Press [Enter↵] or click on OK to begin printing.

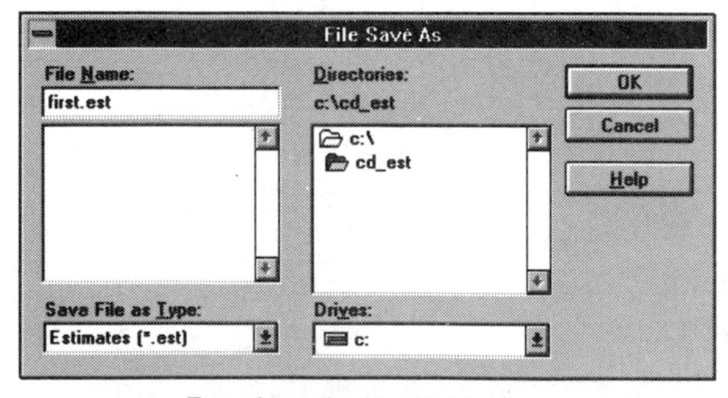

Type the estimate name in the File Name box to assign a file name.

Save Your Estimate to Disk

To store your estimate on the hard disk where it can be re-opened and changed at any time, either:

- Click on the button on the tool bar, -Or-
- Click on File on the menu bar. Then click on Save, -Or-
- Hold the [Ctrl] key down and type the letter S.

The cursor is in the File Name box. Type the name you want to give this estimate, such as FIRST. The name can be up to eight letters and numbers, but don't use symbols or spaces. Press [Enter↵] or click on OK and the estimate is written to disk.

Opening Other Costbooks

The CD Estimator disk comes with six construction cost estimating databases – over 2,500 pages of labor and material costs published in six estimating reference manuals:

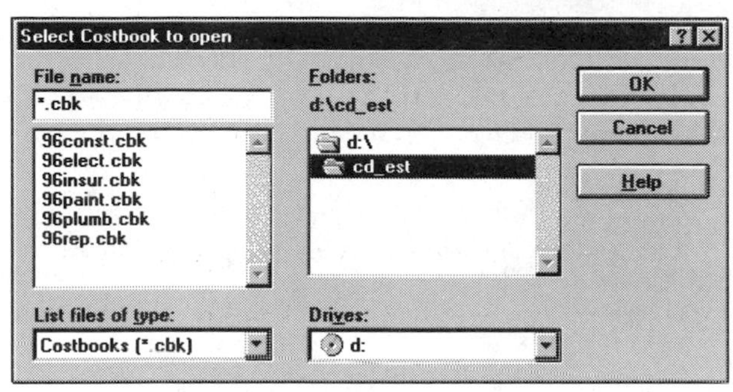

Open the costbook of your choice.

- *National Construction Estimator*
- *National Repair & Remodeling Estimator*
- *National Electrical Estimator*
- *National Plumbing & HVAC Estimator*
- *National Painting Cost Estimator*
- *National Renovation & Insurance Repair Estimator*

All six of these costbooks can be opened on your CD-ROM drive (usually D:\CD_EST). The costbooks installed when you ran SETUP can be opened on your hard drive (usually C:\CD_EST).

To open any of the six costbooks, click on File on the menu bar. Then click on the Open Costbook. Select either the C: or D: drive. Then double click on the costbook of your choice.

To see a list of the costbooks open, click on Window. The name of the current costbook will be checked. Click on any other costbook name to display that costbook. Click on Window, then click on Tile to display all open costbooks and estimates.

Setting Your Default Costbook

Your default costbook opens automatically every time you begin using National Estimator. Save time by making the default costbook the one you use most.

To change your default costbook, click on Utilities on the menu bar. Then click on Options. Next, click on Select Default Costbook. Double click on the costbook of your choice and then click on OK.

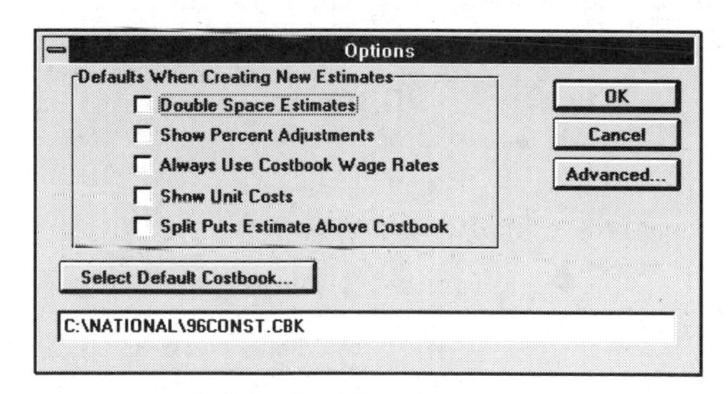

Select the default costbook.

Use National Estimator Help

That completes the basics of National Estimator. You've learned enough to complete most estimates. When you need more information about the fine points, use National Estimator Help. Click on [?] to see Help Contents. Then click on the menu selection of your choice. To print 27 pages of instructions for National Estimator, go to Help Contents. Click on Print All Topics (at the bottom of Help Contents). Click on File on the Help menu bar. Then click on Print Topic.

Click on Print All Topics to print the entire Guide to National Estimator (27 pages).

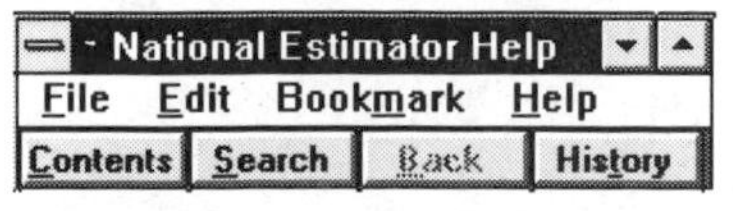

Click on File and then Print Topic.

Free Cost Updates by Modem

If you have a computer running Windows 3.1 or higher, a modem running at 9600 baud or more, and at least 1 megabyte of free space on a hard disk, you're about to discover a better way to collect construction cost estimating data. Costs in your 1996 CD Estimator disk are revised quarterly and made available at no charge on the Contractor's Bulletin Board System (BBS) and Contractor's Home Page on the Internet. You pay only for the phone call. You can get updates to the National Estimator program (TNE.EXE) and the cost data in six construction estimating reference manuals:

- *National Construction Estimator*
- *National Repair & Remodeling Estimator*
- *National Electrical Estimator*
- *National Plumbing & HVAC Estimator*
- *National Painting Cost Estimator*
- *National Renovation & Insurance Repair Estimator*

Contractor's BBS also offers over 20,000 megabytes of shareware programs, 24 hours a day. Many of these programs are of particular interest to professionals in the construction industry. All are free of charge (until your connect time exceeds 45 minutes on any day).

Contractor's Home Page on the Internet

If you have access to the Internet, it may be easier to get the files you need on Contractor's Home Page. The URL is HTTP://ns1.win.net/~contractor/ Instructions beginning on page 175 ("Using Contractor's Home Page on the Internet") describe how to use Contractor's Home Page.

Contractor's BBS

Excalibur Icon

The pages that follow explain how to install the Excalibur communications program, log on to Contractor's BBS, and download revised National Estimator costbooks. Once Excalibur has been installed on your computer, logging on to Contractor's BBS is quick and easy. You'll probably want to download revisions of the National Estimator program (TNE.EXE) and revised costbooks as they become available.

For Windows 3.1 and 3.11 Users

The instructions that follow assume you're using Windows 95. If you're still using Windows 3.1 or 3.11, the procedure is a little different. Craftsman offers a free booklet, *Contractor's BBS and Contractor's Home Page*, that explains how to log on with earlier versions of Windows. This booklet is especially useful if you need some help setting up a modem. Call 619-438-7828 and ask for the free Contractor's BBS booklet.

Testing Your Modem Under Windows 95

When Windows 95 was installed, it should have detected and installed software for your modem. To be sure your modem is installed correctly under Windows 95:

1. Click on Start.
2. Click on Settings.
3. Click on Control Panel.
4. Double click on Modems.
5. Click on Diagnostics.
6. Your modem should be listed beside a Comm port.
7. Click on that Comm port. (Remember the Comm port selected.)
8. Click on More Info...
9. Windows 95 will try to identify any problem with your modem.
10. Remember the name of your modem (if shown).

Installing Excalibur

Contractor's BBS Install

With CD Estimator in your CD drive, click on start. Then:

1. Click on programs
2. Click on Construction Estimating program group
3. Then click on Contractor's BBS Install.

That opens the Welcome To Excalibur! dialog box:

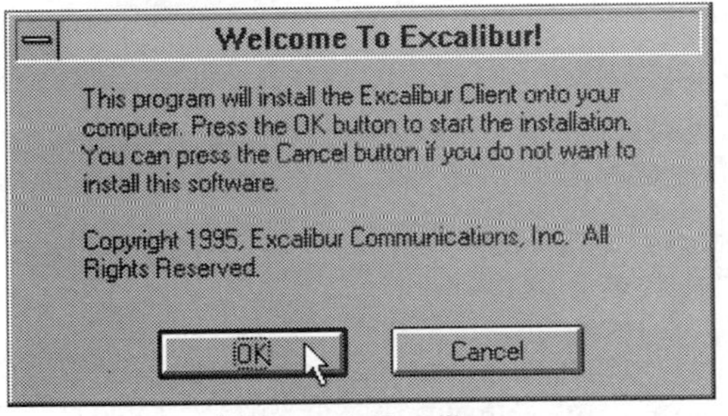

Welcome to Excalibur.

Click on OK to open the Excalibur License Agreement dialog box. Click on OK again to open the Select Destination Directory dialog box.

1. Type the drive letter and directory where you want Excalibur installed. C:\EXCAL is recommended.
2. Click on OK.
3. The Installing dialog box shows installation progress.
4. Installation continues until you see the Install Icons? dialog box.
5. Click on Yes.
6. Click on Construction Estimating in the Select Program Manager Group dialog box. Then click on OK.
7. Installation is complete. Click on OK to continue.
8. Excalibur welcomes you and offers some suggestions for using the program. Click on File and on Exit when you've seen enough.
9. The next step is to get started using Excalibur.

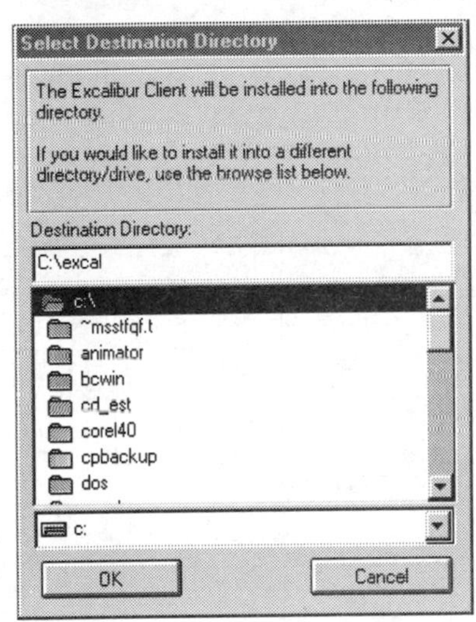

Select the Destination Directory.

Making Connection with Excalibur

Excalibur Client

Click on the Excalibur Client icon if you see it in the Construction Estimating program group. (Otherwise, click on the Start button, click on Programs, click on Construction Estimating, click on Excalibur Client.)

1. Click on OK to fill in user settings. Before using Excalibur the first time, you have to provide some information about yourself and your computer.
2. In the User Information dialog box, fill in your name. Press Tab to move from box to box. Boxes with an asterisk (*) have to be filled in. Other boxes are optional. When complete, click on OK.

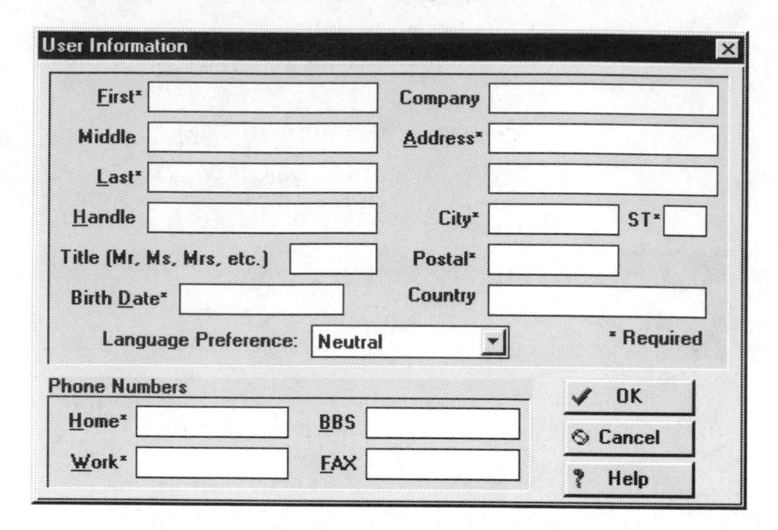

Enter your name, address and phone number. (If the Construction Estimating program group obscures your view, click on the Construction Estimating title bar and drag the box out of the way.)

3. Click on OK to begin filling in modem settings. Most of what's in the Modem Setup and Defaults dialog box will look like Greek to you. But don't worry. You'll use only two boxes here.
4. Click on the down arrow button to the right of the Comm Port box to display communication ports available. Click on the Comm port your computer uses. (If you don't remember, review the section Testing Your Modem Under Windows 95.)
5. Next, click on the Modems button. In the Modem Settings dialog box, click on the up or down arrow to scroll through the list of available modems. Your modem is probably on this list. If not, try one of the generic settings. A little experimenting may be required.

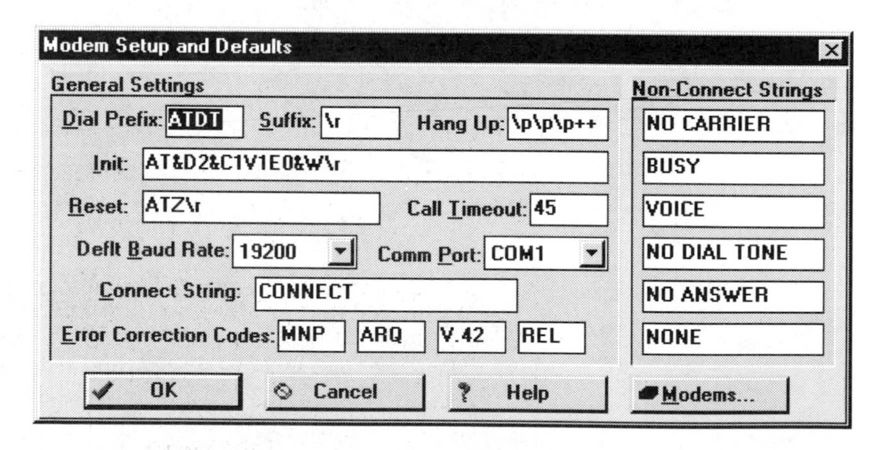

Select your Comm port and modem.

6. When you've selected the modem, click on OK. Click on OK again to save your changes.
7. Then click on System.
8. Click on Dialing Dir...
9. With the highlight on Edit Me, click on Edit Entry.

Edit Directory Entry

System Name: Contractor's BBS
Phone/Internet: 1,319-242-0060
Password: Password
First Name: ROBERT
Middle: L.
Last Name: LEWIS
Baud Rate: 19200
Last On:
Connections: 0
Total Time: 0:00
Reset
Ok | Cancel | Help
Comments
Enter the phone number of the BBS you called to get Excalibur. They would like to have you call back and REALLY get a chance to use their BBS.

Type Contractor's BBS, the phone number and a password.

10. In the Edit Directory Entry dialog box, type Contractor's BBS and press [Tab].
11. In the Phone/Internet box, type 1,319-242-0060 and press [Tab]. (Remember to begin with *70 if you have call waiting or a 9 if you have to dial 9 for an outside line.)
12. In the Password box, type a password you can remember. Then click on OK.
13. In the Dialing Directory dialog box, click on Contractor's BBS. Then click on Dial.

If settings are correct, your modem will begin dialing. Excalibur will display commands sent to the modem and responses received from the modem. After a few seconds, the modem should connect with Contractor's BBS. Occasionally, all lines to Contractor's BBS are busy, especially in the early evening when traffic is heaviest. If the line is busy, try calling back later.

The first time you use Excalibur to communicate with Contractor's BBS, you'll be asked to confirm that you're a new user. Click on Yes. Next, you'll be asked to verify the user information on file. Click on OK when the file data is correct.

Knock on the front door to get admitted to Contractor's BBS.

Welcome to Contractor's BBS.

First Time Log Ons

If this is your first time calling, Excalibur will display a few rules and ask for compliance. Indicate that you plan to comply. From this point, Contractor's BBS guides you gently to the Main Menu. Along the way, you'll have a chance to review information that may be of interest. Take any route you prefer. Eventually, all roads lead to the Main Menu.

Using the Tool Bar Icons

Tool bars and buttons on Contractor's BBS speed and simplify access to information in files and libraries. For example, when viewing the list of libraries, you can click on the preview, search for files or upload buttons.

The Library Tool Bar

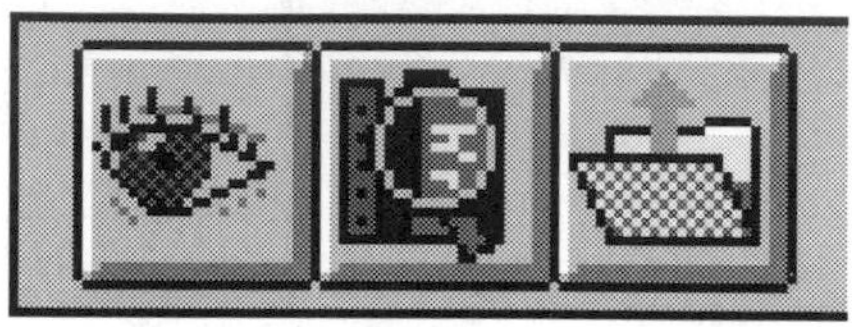
Preview Search Upload

- Click on the library of your choice. Then click on the preview (eye) button to view the library contents.
- Use the magnifying glass button on the tool bar to search for specific files. Type the file name, the search text or the file date and Contractor's BBS will display all files meeting the search criteria.
- The icon with an arrow emerging from a file folder is for uploads. Click on a library. Then click on the upload button to send a file to Contractor's BBS. Uploads are allowed to any library with an asterisk (*) following the library name.

The Files Tool Bar (once you're in a library)

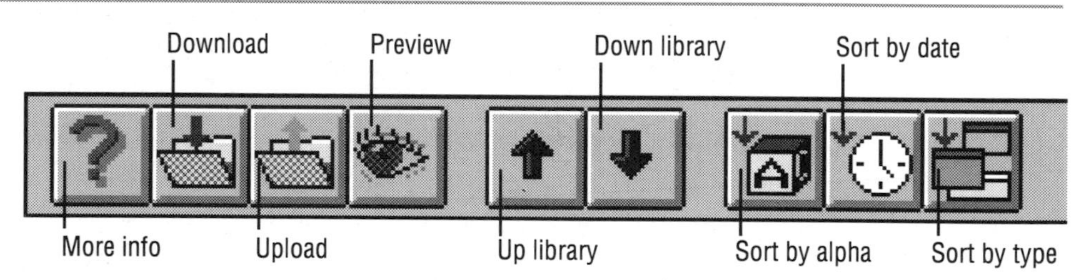

- Click on a file name. Then click on the question mark button to get more information about the file selected.
- Click on a file name. Then click on the download button to have Contractor's BBS send the file selected to your computer.
- Click on the upload button to send a file to Contractor's BBS. In the dialog box, type a description of the file you're sending. Then specify the name and location of the file you're sending.
- Click on a file name. Then click on the preview (eye) button to scan contents of the file. If the file includes graphic images such as GIF files, you'll see a miniature view of that image.

- The Up and Down arrow buttons move you from library to library. This way you don't have to exit the file listing to select a different library.

The last three buttons on the Files tool bar control how file lists are sorted.

1. Click the A letter block to sort alphabetically by the file name.
2. Click on the clock button to sort by file upload date.
3. Click on the button with miniature windows to sort alphabetically by the file type (the last three characters of the file name).

For a brief description of the function of any button, cover the button with your mouse cursor. The button's function appears on the status bar at the bottom of your screen.

Downloading Revised Cost Files

Craftsman Support Forum

Click on the Craftsman logo button anywhere in Contractor's BBS to go to the Craftsman Support Forum. In this forum you'll find event calendars, new and upcoming product information, National Estimator program updates, costbook updates, and a message base specifically for contractors. You can also ask to be placed on the Craftsman Book Company mailing list.

To check for National Estimator cost updates, click on the Upgrades & Updates option. A window with a list box and tool bar lets you select a specific information library. Highlight the library of your choice and click on the preview button on the tool bar.

For example, if you want revised costs for the 1996 *National Electrical Estimator*, click on the 1996 Electrical Estimator library. Then click on the Preview button. You'll see a list of files available, including dates these files were placed on Contractor's BBS, file size, and a brief description. Revised versions of the National Estimator program (TNE.EXE) are posted in the National Estimator Upgrades library.

Revisions to National Estimator cost files are user costbooks and are posted on Contractor's BBS during the third week of the month issued. File names for 1996 are:

Costbook	March Update	June Update	Sept. Update
96const	396const	696const	996const
96rep	396rep	696rep	996rep
96elect	396elect	696elect	996elect
96plumb	396plumb	696plumb	996plumb
96paint	396paint	696paint	996paint
96insur	396insur	696insur	996insur

Updates are cumulative. That means the September update includes all changes made in the March and June updates.

To Download a File

Double click on a file name to see a more detailed description of the contents. To begin transferring a file to your computer, click on the file name. Then click on the download button on the tool bar at the top of the file list box. (The download button has an arrow pointing into a file folder.) During file transfer, you can continue browsing through other files, tagging more files for download, reading messages or even uploading files to Contractor's BBS.

National Estimator cost updates are user costbooks and should be installed on the same directory with other National Estimator files. The recommended directory is C:\CD_EST. Cost updates for National Estimator are not compressed and can be used immediately. To open the user costbook you've downloaded (such as 396CONST.UBK), start National Estimator. Be sure the appropriate costbook is open (such as 96CONST.CBK). Click in that costbook. Then click on File. Click on Switch User Costbook. Double click on the name of the user costbook you've downloaded (such as 396CONST.UBK). After a few seconds, the name of that user costbook will appear on the status bar at the bottom right of your screen. Scroll through the costbook and you'll notice that the revised lines are in red type

What Are .ZIP Files?

Most of the files on Contractor's BBS have a file type of .ZIP. That's your clue that the file has been compressed and has to be decompressed before use. A single ZIP file can contain many files — even many files on many subdirectories. That's handy, because all files needed to run a program can be accumulated and compressed into a single .ZIP file. When unzipped, the files and directories expand to their original form and size. Unzipping is done with a program from PKWARE called PKUNZIP. Here's how to unzip a file:

In the Craftsman Support Forum under "Upgrades and Updates" you'll find a library titled Miscellaneous Utilities. Click on the library name. One of the files in that library is PKZ204G.EXE. That's the latest version of PKUNZIP. Click on that file name. Then click on the download button. PKUNZIP will be sent to your computer. While the file is being sent, find another file you need, such as TNE.ZIP (the National Estimator program). Click on that file name. Click on the download button again and that file will also be sent to your computer.

When file transfer is complete, leave Contractor's BBS by clicking on the exit button. Exit Windows. Change to the directory Excalibur uses when downloading files. The suggested download directory is C:\CD_EST. The two files you downloaded should be in that directory. Look for PKZ204G.EXE and TNE.ZIP. Then type PKUNZIP TNE.ZIP and press Enter. If the TNE program is already installed on that directory, PKUNZIP will ask permission to overwrite with the new version. Type Y and press Enter to overwrite the older file.

Use the same procedure to download and unzip any of the thousands of programs on Contractor's BBS.

The Next Time You Log On

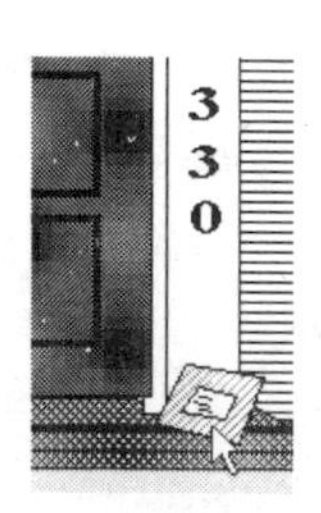

New versions of the Excalibur program are released several times a year. Occasionally a new release will require revision of your Excalibur program file. In that case, Contractor's BBS won't let you in the front door until the new program is installed. You'll be asked to pick up a package waiting on the front porch. Click on OK. Then click on your package. Contractor's BBS will send the new version of Excalibur to your computer. Allow a few minutes to complete the file transfer. (New releases of Excalibur are usually much smaller than the original program.)

When file transfer is complete, exit Contractor's BBS and the Excalibur terminal program. The next step is to run the program you just received. Open File Manager and switch to the download directory you requested. In that directory, find the file you just downloaded. The file name will start with the

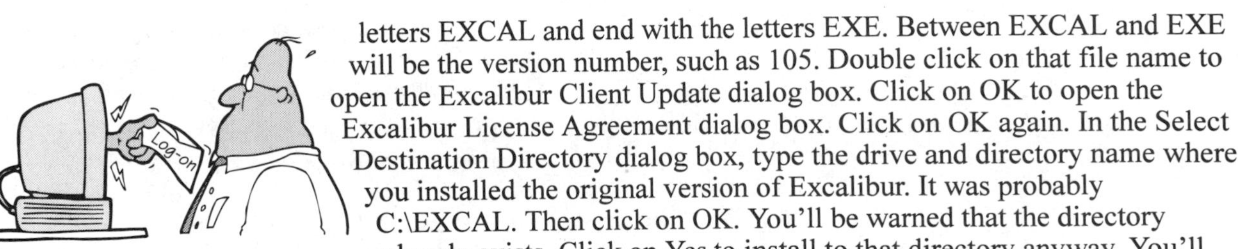

letters EXCAL and end with the letters EXE. Between EXCAL and EXE will be the version number, such as 105. Double click on that file name to open the Excalibur Client Update dialog box. Click on OK to open the Excalibur License Agreement dialog box. Click on OK again. In the Select Destination Directory dialog box, type the drive and directory name where you installed the original version of Excalibur. It was probably C:\EXCAL. Then click on OK. You'll be warned that the directory already exists. Click on Yes to install to that directory anyway. You'll see a dialog box displaying download progress. When asked if you want to add icons to Program Manager, click on No. Next, you'll see some notes on Excalibur. Glance over those notes. Then click on File and Exit.

When overwriting is done, you can delete the file just received. Then restart Windows. At the Windows Program Manager, double click on the Excalibur icon. This time you should have no trouble logging on to Contractor's BBS.

Exploring Contractor's BBS

The Craftsman Support Forum is only a small part of Contractor's BBS. There's lots more to explore. You can leave messages for other BBS users or ask for assistance. Whether you call to get the latest cost updates, to download files, or just to satisfy your curiosity, we hope you can make good use of Contractor's BBS.

On-line Assistance (Paging the SysOp)

Unfortunately, the SysOp (System Operator) is rarely available for chatting. If you have a request or comment, leave an E-Mail message for the SysOp. The reply will probably be waiting the next time you log on. If you need help using Contractor's BBS, call Craftsman Technical Support at 619-438-7828 from 8:00 am to 5:00 pm Pacific time.

Sound Card Compatibility

If your computer has a Windows-compatible sound system, consider downloading the sound file library for Contractor's BBS. Sound files are available from the Main Menu. Click on Files (the filing cabinet). Click on the BBS Support Files button. In the support files library, click on WAVFILES.EXE. Click on the download button to begin downloading. When the download is finished, copy WAVFILES.EXE to your Windows directory. In File Manager or the Windows Explorer, double click on C:\WINDOWS\WAVFILES.EXE to decompress the library. (Once decompressed, you can erase WAVFILE.EXE.) The next time you log on to Contractor's BBS, you'll hear instructions and comments.

http://ns1.win.net/~contractor

Using Contractor's Home Page on the Internet

If you're an experienced Web surfer, point your URL at:

http://ns1.win.net/~contractor

If you're new to the Internet, instructions that follow will be helpful. These instructions assume that you've installed a Web browser (such as Netscape) and have an account with an Internet provider.

1. Start by invoking your Web browser. In the Location or Go to box, type: http://ns1.win.net/~contractor. Then press Enter↵.

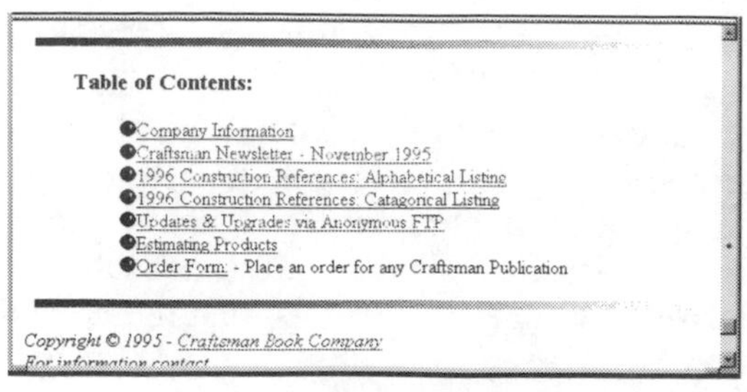

Click on Updates & Upgrades

2. In a few seconds (depending on the volume and capacity of your Internet provider), Contractor's Home Page will open.
3. If your browser has bookmarks, save the current location as a bookmark. That makes it easy to connect next time. Click on Bookmarks. Then click on Add a Bookmark. The next time you want to use Contractor's Home Page, simply click on Bookmarks and then on Contractor's Home Page.
4. Scroll down the page by clicking on the down arrow at the lower right corner.
5. Click on underlined text to branch to what the underlined text describes. For example, to download the latest version of the National Estimator program, click on "Updates and Upgrades via Anonymous FTP."

Files available for downloading include:

cfandc/	Construction Forms & Contracts
news/	Craftsman Newsletters and Reviews
const/	National Construction Estimator
elect/	National Electrical Estimator
paint/	National Painting Cost Estimator
plumb/	National Plumbing & HVAC Estimator
rep/	National Repair & Remodeling Estimator
insur/	National Renovation & Insurance Repair Estimator
tne/	National Estimator program (TNE)
utility/	Includes the PKUNZIP utility

Updates for annual cost estimating books are user costbooks (UBK files). See the section "Downloading Revised Cost Files" on page 173 for instructions on installing and using these files.

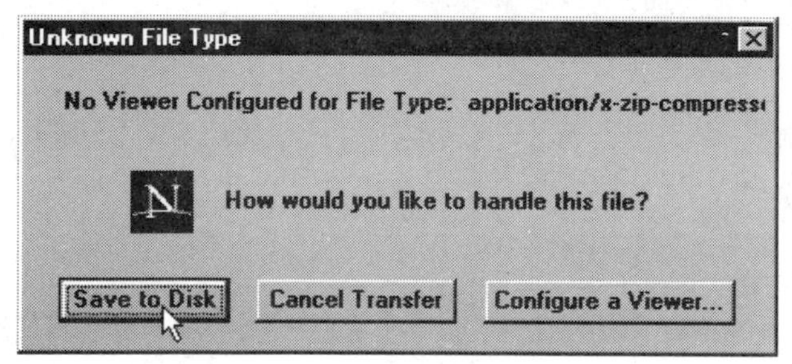

Save the file to disk.

To download the latest release of the National Estimator program, click on TNE/. Then click on TNE.ZIP. This will either begin the download or open a dialog box asking how you want to handle the file. Click on the Save to Disk button.

A dialog box may open asking for information about the download directory. Select the same directory where National Estimator is installed (probably C:\CD_EST). To change a directory, double click on the file folder to the left of C:\. That produces an alphabetical list of directories on your C drive. Double click on CD_EST to select that directory. Then click on OK to start the download.

Don't click on anything else until the file transfer is complete. Any interruption will cancel the download. Download progress is displayed in the status bar at the bottom of your Netscape window.

Any file that ends in ZIP can't be used until it's unzipped. See the section "What Are .ZIP Files?" on page 174 for instructions on how to unzip TNE.ZIP. The file PKZ204G.EXE in the UTILITY/ directory is made to unzip these zipped files.

Nearly every home page on the Internet is under construction. Contractor's Home Page is no exception. Do a little browsing and you'll probably discover lots more of interest.

Chapter 8
Cost Recording

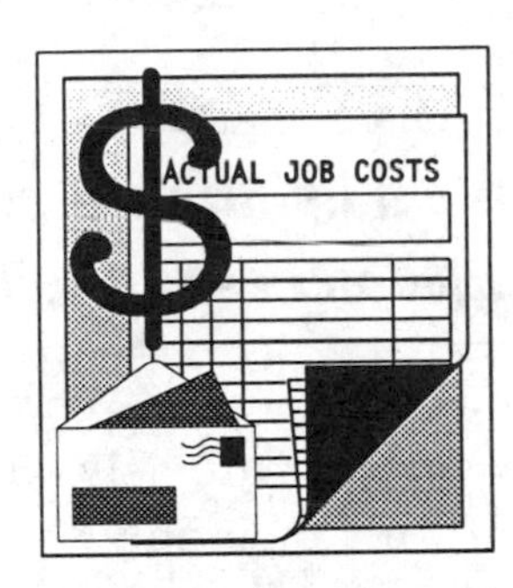

Through the first seven chapters of this book, I've explained how to prepare construction cost estimates: visit the site, review the plans, do the quantity take-off, and price the work. That completes the estimating process. But whether you estimate by hand or with a computer, your work isn't done when you finish the estimate. You still have to sell the job.

But even after the job is sold, your work isn't done. And what's left isn't all construction work. In the rest of this book, I'll describe how the estimate is part of a cycle that helps you produce even more effective estimates and increase your profit.

What's left after the estimate?

You have four tasks left.

1) Control the work so costs come in as estimated.

2) Identify costs that weren't estimated correctly.

3) Record actual costs for use on future estimates.

4) Figure out if the company made or lost money on the job.

I'll cover each of these topics in this chapter. We can broadly define most of this work as "cost recording." But I'm sure you'll see that there's more to it than simply recording costs. Someone has to compare actual costs with the estimates, change what has to be changed, and make sure today's lessons become tomorrow's tools.

Why Keep Cost Records?

Construction is a complex manufacturing process. A builder who has little or no payroll is the best informed. He knows both the exact cost of every part of the work and how long it took to complete every task — he either did it, or he watched it done. That's not possible in a larger company with many work crews, several supervisors and many subcontractors. Someone has to collect cost records (installation manhours, material invoices, subcontractor statements) and assemble that information into a useful form. The person charged with that task is usually the project manager or the estimator who prepared the bid for the job. Sometimes that's the same person.

Reliable cost information is essential

There's no way to make a living as a construction cost estimator if you don't have access to reliable cost data. The more information you collect, and the better that information is, the more effective you'll be at forecasting costs. If you want consistently reliable estimates, you have to collect cost data from completed projects.

"There's no way to make a living as a construction cost estimator if you don't have access to reliable cost data."

Of course, your cost records will be extremely valuable when you prepare future estimates. But that's not their only value. Cost records will help you identify wasted materials, wasted time, wasted supplies, poor supervision, poor job layout, and extra costs that should be charged to the property owner.

Fortunately, cost-keeping isn't hard. Any estimator can do it, and all of them should. Follow the suggestions here, use the forms we recommend, and you'll soon be accumulating valuable construction cost data — with very little extra effort, and in a minimum of time.

If you already have a cost-keeping system that's working well, congratulations. Keep using it and skip to the next chapter. If, like many smaller construction companies, you aren't using a cost-keeping system, keep reading. This chapter has the forms and information you need to set up a good cost-recording system for your company.

Cost-keeping Described

Your cost-keeping system should contain the cost for every unit of completed work. You'll collect information on each task, the time it took, and any special conditions that affected completion of the job. The goal is simple: to identify actual costs, both per unit installed and for the entire project.

Sure, it takes time to collect this information. I won't deny that. Most workers and supervisors don't like any type of paperwork, no matter how streamlined and simplified. But the value of this information in your hands will be far greater than the burden it places on your field crews to produce it. You won't make yourself popular when you require extra paperwork, but you *will* become a better estimator. And that's good for everyone who works for your company.

Cost-keeping vs. bookkeeping

While cost-keeping is more like bookkeeping than estimating, bookkeeping and cost-keeping aren't the same. Both record where the money went. But cost-keeping helps you find unit costs (per square foot of wallboard or cubic yard of concrete). Bookkeeping deals with profit and loss. Cost-keeping is an engineering function. Bookkeeping is an accounting function.

Essentials of a Cost System

Regardless of the size of a construction company, any effective cost-keeping system has to be:

Four qualities of effective cost-keeping

1) Reliable
2) Simple
3) Inexpensive
4) Flexible enough to handle any type of work

The system I'm going to describe here meets these standards.

All project costs are either labor, material, equipment, or project overhead. Your cost system has to identify each of these costs for everything on your estimate, from the time you present the bid until you turn the finished building over to the owner.

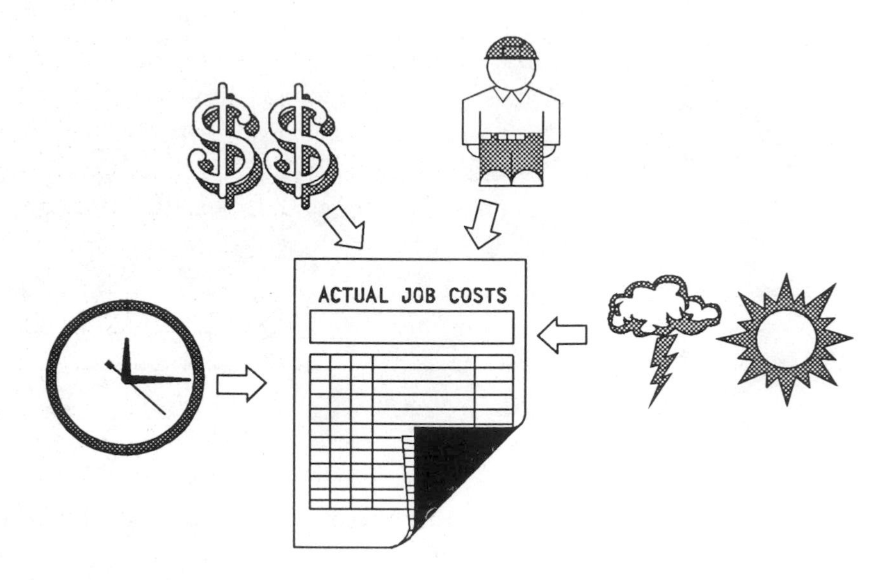

Cost Analysis

It's not enough just to keep track of the *total* labor, *total* material, *total* equipment and *total* overhead cost on a job. True, it's important to compare actual totals with estimated totals. But that doesn't tell you where any estimating mistakes were. You need a breakdown of actual costs per unit, such as per cubic yard of concrete foundation, or per square foot of floor. And you also need to know costs per unit for material, labor, equipment and project overhead. In short, you have to keep cost records at the same level of detail and with the same descriptions that you used when you estimated the job.

An elaborate system isn't necessary

For simple estimates - A contractor who subs out nearly all the work on a job probably doesn't need an elaborate cost classification system. It's easy to keep track of actual costs. Just make a photocopy of your estimate. Write "Actual Costs" at the top of the form in red ink. Then list (in red) the actual subcontract, material, labor, equipment, or overhead cost beside each estimated cost. When you've finished the job and written all the actual costs in red, compare estimated costs with actual costs. Where actual costs are higher, find out why.

If you use *National Estimator* to bid jobs, it's easy to create a record of actual job costs. Click on Save As. In the Save File As Type box, select Text File for export to word processing. For export to Lotus 1-2-3, select Comma Separated. For export to Excel, select Tab Separated. Then click on OK to save the file. Next, open the estimate in your favorite word processing or spreadsheet program. Change estimated manhours to actual manhours and estimated materials costs to actual job costs. Change the heading from "Construction Estimate" to "Actual Job Costs" and print a copy for your file. The format will be nearly the same as your original estimate. That makes it easy to compare the two documents. A collection of cost records like this will be your most useful estimating reference.

More complex estimates - If your estimate is many pages long, obviously you'll need a good classification system to keep track of manhours and costs for each part of the job. The classification system has to be flexible enough to record actual costs for all the construction tasks your company handles. It has to be detailed enough to have a category for each item listed in your estimate.

Use the CSI Masterformat classification system

You could make up your own classification system, but I think a good one already exists. As we discussed in Chapter 3, the CSI Masterformat divides construction tasks into about 10,000 categories, under 16 major divisions. That's more detail than almost any contractor needs. If you do only masonry, for example, you'll use only the masonry classifications. In the Masterformat system, each category has its own five-digit number. For example, Section 04, Masonry, includes the following numbered categories:

04100 Mortar

04150 Masonry Accessories

04200 Unit Masonry

04210 Clay Unit Masonry

04220 Concrete Unit Masonry

04230 Reinforced Unit Masonry

04235 Pre-assembled Masonry Panel

04240 Non-reinforced Masonry

04270 Glass Unit Masonry

04280 Gypsum Unit Masonry

04290 Adobe Unit Masonry

04400 Stone

04410 Rough Stone

04420 Cut Stone

04440 Flagstone

04450 Stone Veneer

04455 Marble

04460 Limestone

04465 Granite

04470 Sandstone

04475 Slate

04500 Masonry Cleaning

04550 Refractories

04600 Corrosion-Resistant Masonry

04700 Simulated Masonry

You can get a copy of Masterformat from Construction Specifications Institute. The address is on page 39 of this book.

To get the detail you need, you can assign an intermediate digit to identify the type of unit masonry involved. For example, under 04210, Clay Unit Masonry:

04210 Clay Unit Masonry

04211 Common Brick

04212 Face Brick

04213 Pavers

04214 Fire Brick

04215 Veneer Strips

04216 Cap Brick

04220 Concrete Unit Masonry

Then add a letter M, L, E, or O to indicate Material, Labor, Equipment or Overhead cost. For example, use 04211M for material cost of common brick and 04211L for labor cost for common brick. If you need to identify where the brick was used, add a dash and a short description. For example "04211L-Walls" means labor cost for brick walls and "04211L-Piers" means labor cost for brick piers.

Once you start work on a project, write down job costs as you receive the invoices or pay the bills. Assign a classification number to each cost, and record the payment amount. It won't be hard to list material, equipment and project overhead costs. But it's harder to keep track of labor costs. That's what I'll cover in most of the rest of this chapter.

Cost Records Check Current Jobs

As work progresses, your records should show the cost of materials ordered and delivered so far, and how many manhours have been spent to date. Regular comparisons of actual with estimated costs will help you identify potential problems while there's still time to make corrections.

For example, suppose your estimate for concrete block on a job was $4,020. The breakdown was $2,200 for materials, $1,200 for labor and $620 for overhead (mostly supervision) and equipment. Your cost record so far shows charges of $3,900 and the foreman reports that the job won't be finished for two more days. If it takes two more days, the labor cost will be at least $1,500, fully 25 percent more than originally estimated. Obviously, costs are running higher than planned. Something is wrong, either with your estimate or with the supervision or with the tradesmen or with the job. What do you do?

Don't ignore any problems

Here the worst thing you could do: ignore the problem. The next worst thing is to forget about it until the job is done. The best option is to find out what went wrong while the crew and supervisor are still working. Start asking questions. See for yourself why progress is so slow. Talk to the crew. With a little luck and some subtle persuasion, you might even be able to cut one of those two extra days out of the job.

Classifying Labor Costs

To find accurate labor costs for each item in your estimate, you have to know how long it took to do each task. You have to know how many hours each crew worked, and what they did during that time. If you think it sounds simple to collect that information, it's because you've never tried it.

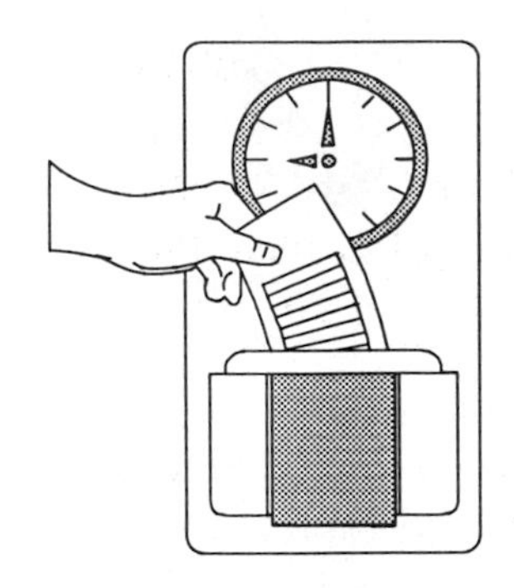

Here's how to make the job as easy and painless as possible: Make cost recording a part of your payroll system. Every company with a payroll keeps employee time records. The method varies from company to company. But usually each worker submits a daily time card which shows the time he or she worked. In a larger company, a foreman or timekeeper may keep track of hours. A supervisor should always check the time cards before they go to the payroll clerk who prepares the paychecks.

To record manhours per task, get either the supervisor, timekeeper or worker to record both the total work hours, and the approximate time spent on each task. Absolute precision isn't necessary. Records to the nearest 30 minutes will usually be good enough. But you'll have to do some coaching to get the detail you need. For example, "setting forms" may not be specific enough. You need to know if these were foundation forms, retaining-wall forms or sidewalk forms.

Using the Time Slip

Figure 8-1 shows a simple time slip that would work for both payroll and manhour recording purposes. Print a supply of these on heavy paper or card stock, and bind them into pads. Have your workers complete the time cards daily, while the work is still fresh in their minds. After the supervisor verifies the hours, he sends the time cards to the payroll clerk who writes the checks. Then, the payroll clerk returns the time cards to the estimator/project manager, who records the actual manhours for each task.

Insist that all time cards are completely filled out

Don't accept time slips unless they show the start and end time for each task. Be sure your supervisors check the time cards, and insist that the workers supply all the necessary information. The most common problem will be in the "Work Description" column. Of course, you'll be sure your supervisor has a copy of the original estimate. If there's a question about what tasks to list, the estimate will answer it. What you need is the actual manhours for each manhour estimate listed.

My experience is that crew members seldom work on more than six or eight separate tasks in a day. In many cases, tradesmen work on a single task all day, such as "framing walls" or "shingling roof." You'll seldom need to use more than one time slip for any worker on any day. A single slip should have more than enough room to record six or eight separate tasks. But your supervisors will probably work on several jobs each day. If so, be sure those supervisors turn in a separate time slip for each job they visit each day. Any task that takes more than about 30 minutes is worth recording as a separate item on the supervisor's time slip.

Avoid collecting too much information

Avoid the temptation to collect *too* much information on the time slips. I've seen daily time slips that require workers to list not only the total time they spent on each task, but the amount of work they got done, in square feet, for instance. I don't think that's necessary. You know how many units of work have to be done. The worker doesn't have to know, and may have trouble calculating just how much work he or she actually did. Just ask for

Skyline Builders
Daily Time Slip

Work done by ______________________ Payroll # __________

Job name ______________________ Date __________

Work description	Comment	Start Time	End Time	Total Time	Code #	Adjust	Adjust Total
			Totals				

Figure 8-1 Daily time slip

a brief description of what the workers did, what time they started, and when they finished. Then you, your supervisor, or your estimator/project manager fills out the shaded part of the form: Total Time, Code Number, Adjustment, and the Adjusted Total.

The Code Number column makes it easy to collect actual costs, especially if you use a computer for cost recording. Write in this column the same account number you used when estimating this work. It's also the same account number you'll use on the Actual Job Cost form later in this chapter.

Adjusting manhours - Here's how to use the last four columns on the time slip:

The "Total Time" column on every card must reflect all the hours the worker expects to be paid for. That's elementary. Unfortunately, not all time spent "on the clock" will be *productive* time. You'll get many time

Skyline Builders
Daily Time Slip

Work done by Jack Jones | Payroll # 14

Job name Wilson St. Apts. | Date 3/4/93

Work description	Comment	Start Time	End Time	Total Time	Code #	Adjust	Adjust Total
Bottom Plates		7:30	9:45	2.25		1.143	2.57
Wall Framing		9:45	11:30	1.75		1.143	2.00
Wall Framing		12:00	1:30	1.50		1.143	1.71
Waiting	Rain	1:30	2:30	1.00		–	—
Wall Framing		2:30	4:00	1.50		1.143	1.71
			Totals	8.00		1.143	8.00

Figure 8-2 Completed time slip

slips which include tasks that are clearly non-productive and aren't in your estimate. For example, you'll see time cards with tasks such as "went to lumberyard," or "waiting for power hookup."

Two ways to handle non-productive labor

There are two ways to handle non-productive labor such as this. First, you could charge that time to overhead. Multiply the time by the hourly rate (including the labor burden) and charge the total cost to job overhead. But that method will usually inflate overhead expense more than most estimators (or company owners) are willing to tolerate.

A better choice is to distribute non-productive time proportionally among productive tasks. For example, suppose your carpenter turned in a time slip like Jack's in Figure 8-2. His productive time for March 4, 1993, was 7 hours. Jack also spent 1 hour waiting for the rain to stop. Divide 8 by 7 (total time divided by productive time) to get 1.143. That's your adjustment factor. Write 1.143 in the "Adjust" column for each of the

Minutes	Hr/100	Minutes	Hr/100	Minutes	Hr/100	Minutes	Hr/100
1	2	16	27	31	52	46	77
2	3	17	28	32	53	47	78
3	5	18	30	33	55	48	80
4	7	19	32	34	57	49	82
5	8	20	33	35	58	50	83
6	10	21	35	36	60	51	85
7	12	22	37	37	62	52	87
8	13	23	38	38	63	53	88
9	15	24	40	39	65	54	90
10	17	25	42	40	67	55	92
11	18	26	43	41	68	56	93
12	20	27	45	42	70	57	95
13	22	28	47	43	72	58	97
14	23	29	48	44	73	59	98
15	25	30	50	45	75	60	100

Figure 8-3 Minutes to hundredths of an hour

productive tasks. Then multiply the time spent on each productive task by 1.143. Notice you ignore the non-productive time spent waiting for the rain to stop. The adjusted total for productive time is also 8 hours.

Also note on Figure 8-2 that I've converted hours and minutes to hundredths of an hour in the Total Time column. That makes it easier to add and multiply the times in that column. Figure 8-3 is a minutes to hundredths-of-an-hour conversion table you can use.

Use time slips to increase productivity

I've found that these time slips increase productivity rather than reduce it. Even when there's no supervisor on the job, most workers will be more productive if they have to account for time they spend on each task. No tradesman is going to write in 30 minutes for drinking coffee or talking about last night's game. But many will spend more than 30 minutes doing exactly that if they're not accountable for their time. Because time slips increase accountability, they reduce wasted time.

But there's no advantage to making workers feel pressured to produce or meet a quota. Anyone can get a job done in record time by doing sub-standard work. That's not the purpose of time recording. We all want to reduce wasted time. But we all know that sometimes jobs take longer than we expect. Anyone who's tried to back a corroded nut off the underside of a wash basin in a vanity cabinet knows it's possible to work hard for three-quarters of an hour doing what would normally be a 30-second chore.

Expect the unexpected - Delays and surprises are common in all types of construction: The owner shows up on the job and insists that the painter wire-brush some paint splatters off the concrete. Or a drill bit breaks and you have to go back to the shop for another one. That's the nature of construction. It's manufacturing, but it's never done in a factory under easily-predicted, controlled conditions. You'll meet the unexpected on almost every job. That's the purpose of the *Comment* column. You use it to record the unexpected.

Under *Comment*, encourage workers to write anything unusual that the estimator should know about the job. Employees get the point of that right away. They'll list excuses: It was too hot or too cold or too windy or it rained or the subs didn't show up or they ran out of nails or the saw broke or the lumber was shoddy or the plans were wrong or . . . Well, you get the idea. Many tradesmen will be more anxious to fill in the *Comment* column than any other part of the time card. That's great, because I like to see those excuses. They help me understand what went wrong on the job and what could be done to improve future jobs — both my estimates of them, and the work itself. I encourage workers to unload their frustration in the *Comment* column. Let them use the back of the form if necessary. I read it all and love to see it.

Using Cost Data

Figure 8-4 shows the form I use to compile actual costs for a job. Notice that this form looks a lot like an estimate. The major difference is the two extra columns:

Differences in the Actual Job Cost form

1) A *Code #* column for the Masterformat number (if you choose to classify costs by number)

2) A column showing manhours per unit

Use the *Quantity & Description* column to describe any unusual conditions that affected productivity.

Costs per unit change as labor rates change. Labor costs per square foot or per cubic yard on an old estimate won't apply any longer if hourly pay rates have changed. But the labor hours per unit should be about the same year after year. That makes the *Manhours* column on Figure 8-4 the most useful when you estimate the cost of later jobs.

Skyline Builders
Actual Job Costs

Job name ______________________ Estimate # ______________________

Start date ______________________ Finish date ______________________

Quantity & Description	Unit	Man-hours	Code #	Material Cost	Labor Cost	Equip-ment	Total Cost

Figure 8-4 Actual job costs

Compiling Actual Job Costs

I start a file folder for each job when I get the first bills (or time cards) for a new job. I file all the bills and time cards for that job in the folder. If I get a bill for materials delivered to three different jobs, I make a note on the bill showing the cost of material delivered to each job. Then I make copies and file a copy in each of the three job folders.

Several times a month I go through each job folder and sort the bills and time slips into the same order as I estimated the work. For a larger job, this sorting goes faster if I note the Masterformat number assigned to the work we're being charged for. Every cost, whether labor, material or equipment has a Masterformat number that I've assigned in my original estimate. I make sure that correct category numbers are listed on the Daily Time Slips. When I pay a bill for materials, I note beside each item on the invoice the Masterformat account number that identifies those materials on my estimate. I make sure the correct account number appears on our copies of equipment rental tickets.

Then I record actual labor, material and equipment costs on a form like Figure 8-4. For an eight-page estimate, it takes about an hour to an hour and a half to record actual costs. That's assuming, of course, that subs and material suppliers described their charges correctly and my workers listed their time accurately.

"The cost information in that binder is both my most used and most useful estimating reference."

When I'm done, I have a good picture of actual costs and an excellent guide to use when I estimate similar jobs in the future. Once I've compared estimated and actual costs (and learned what I can about my mistakes), I file the Actual Job Costs form behind the estimate in my three-ring binder.

The cost information in that binder is both my most used and most useful estimating reference. No company will ever publish construction cost data that's more accurate for the work my company handles and for the tradesmen and supervisors my company employs. Without question, it's worth many times the trouble and effort it takes to keep these cost records.

Maybe most important of all, the Actual Job Costs form shows at a glance whether I made or lost money on the job. I know it sounds funny, but some contractors never find out if they made or lost money on a job.

They have no record of actual costs so they aren't sure which jobs (if any) were profitable and which jobs were losers. That's pure foolishness, in my opinion. How can you concentrate on what you do best if you haven't a clue about what you did right? The real pros in this business would never make that mistake. You shouldn't either.

This Is Both Beginning and End

This chapter is a transition between the two sections of this book. The first seven chapters covered the basics of construction cost estimating. Chapters 9 through 12 cover more advanced topics. They'll help you bid smarter and more successfully, no matter what type of work your company handles.

Chapter 9
Planning Overhead

In Chapter 2, I talked about business planning and asked you to develop a plan for your own company. I hope you at least sketched out some goals for the long term — the next five to seven years. In this chapter, we'll begin using those goals by planning for the coming year.

Your one-year plan and budget should be much more specific than your long-term plan. The one-year plan should have two parts. The first is your overhead budget: what you plan to spend on overhead in the coming year. We'll use that figure to compute the correct amount to add for overhead on your estimates. The second part of your plan should be a list of changes the company must make to stay on track with the long-range business plan. More about that later. For now, let's concentrate on the first part of your plan, the overhead expense for the next year.

Company overhead vs. project overhead

In Chapter 4 I divided overhead into two categories, company overhead and project overhead. You'll remember that company overhead is business expense (like phone, light, heat, and accounting fees). Company overhead would go on about the same even if all crews were idle next week. Project overhead is different. It's job site costs like permits, rental costs and non-productive labor (including supervision). If all crews stopped work next week, project overhead would drop to zero.

All project overhead for a job should be accounted for item by item on your estimate — such as "Building Permit Fee," or "Job Shack Rental," or "Cleanup." You'll list company overhead as a single item, "Overhead" and almost certainly calculate it as a percentage of job cost.

The overhead I'm talking about in this chapter and the next one is company overhead. It's much harder to estimate than job overhead. I think estimators make more mistakes estimating overhead than anywhere else. Here's why. You're pretty sure about the cost of 2 x 4s and you know what concrete runs per cubic yard. If you're not sure, you just pick up the phone and get a quote. But do you really know your overhead expense? Who do you call to get a quote on your overhead? Nobody knows except you. And you'll never know unless you sit down and figure it out. Fortunately, that's not too hard. Unfortunately, many contractors never do it.

If you know how to figure overhead and have been doing it all along, good. Follow my discussion in this chapter to see if what I recommend is the same as what you've been doing. If you've never figured company overhead and planned overhead for the coming year, that's OK. You're probably in the majority. But now's the time to begin.

The next time you're at a contractors' meeting or convention, ask a few builders how much they spend on overhead. I'll bet you get more blank stares and off-the-wall guesses than straight answers. That's because most builders don't *know* how much they spend on overhead. Many routinely add 10 or 15 percent to their bids to cover overhead. But how many base that on a calculated figure? Not many. At least not many small contractors. The big boys, engineering and construction contractors handling million-dollar jobs, know their overhead to three decimal places. But very few smaller contractors do. And that's a pity, for two reasons.

First, overhead is very important. For most contractors and for most types of work, overhead will be one of your biggest expenses — probably more than concrete or more than lumber or more than drywall. You wouldn't guess about any of these costs. So why guess about overhead?

Second, knowing your overhead cost is the first step in controlling it. Later in this chapter I'll recommend that you budget overhead. First you figure out how much you're spending on overhead items. Then you do what you must to keep overhead expense in line with revenue. For most contractors, that's an essential key to making money in this business.

Start With Last Year

Your best guide to overhead expense for the next 12 months is your actual overhead expense for the last 12 months. So you start budgeting expense for the next year by finding your actual expense for the last year. Use Figure 9-1 to show how much you spent on overhead over the last 12 months.

Look at the major headings in Figure 9-1. Several of them are broken down into detailed expense categories. When you're ready to fill out this form, go back through your tax returns, check stubs, receipts and invoices

Description	Last 12 mo.	Chg + or -	Projected	Description	Last 12 mo.	Chg + or -	Projected
Accounting				**Office expenses**			
Collections				Building maintenance			
Advertising				Equipment repairs			
Amortization				Postage			
Bad debts				Printing			
Depreciation				Rent			
Donations				Security			
Dues and subscriptions				Supplies			
Entertainment				Telephone			
Equipment (indirect)				Utilities			
Fuel				**Personnel (indirect)**			
Maintenance				Payroll			
Rental				Pensions			
Repairs				Profit sharing			
Small tools				Union dues			
Insurance				**Taxes**			
Liability				Payroll (indirect)			
Property				Property			
Interest				**Travel**			
Legal				**Vehicles**			
Licenses and fees, business				Fuel			
Miscellaneous				Insurance			
				Registration			
				Repair and maintenance			
Totals				**Totals**			

Figure 9-1 Overhead recap form

to determine how much you spent last year for every item on that list. Be as thorough as you can. If you don't have an exact figure, your best guess will be good enough. List actual overhead costs for your business under the column heading *Last 12 Months.* Then total that column.

Your figures for next year probably won't be exactly the same as those for last year. Expenses will change in at least a few categories. If your company is growing, expenses will change in many categories. It's harder to figure the changes for next year than it is to add up last year's costs. But you should be able to make some informed guesses. And remember, a guess is better than no estimate at all. Enter these changes under the column *Chg + or -*. If you expect an expense to change, write in the amount, either in dollars or as a percentage of last year's figure.

For example, suppose you decide it's time to hire a part-time assistant estimator. Add that cost to your payroll expense. Or maybe you want to buy a new computer to handle estimating. The cost of the computer itself isn't the only (or maybe even the largest) expense. Computers are so cheap now that many construction companies can buy one out of petty cash. The expensive part is the time it takes to learn the system. Here are some questions you'll need to ask:

The hidden costs of owning a computer

- How much will you spend on computer, printer and monitor?
- Will you finance it? How much will you pay in interest?
- How much training will employees need?
- How much staff time will it take to switch over to the new system?
- What's the annual maintenance cost for computer, printer and monitor?
- How much will you spend for supplies?

I'll guess that this change to estimating with a computer would add about $6,500 to overhead for next year: $5,000 for hardware and software (computer, printer, monitor, computer programs and supplies), $1,000 for labor time lost in training and $500 for maintenance during the first year. Of course, you can expect the computer to more than repay this investment over the next few years. But there will still be extra overhead expenses during the first year.

Continue to list extra expenses you anticipate during the coming year. Most of these will be changes you'll make to stay on track with your long-term business plan. Then list any savings you can expect. For example, maybe you'll pay off the note on a company pickup truck in three months. With no payments due after then, interest expense will be lower during the next year, unless you take out another loan.

Suppose my long-term plan is to open a branch office in another city where prospects for construction volume are better. Next year I plan to bid and win a job there as a test of that market. That test will cost me more overhead money to bid and build, beyond what's required for my usual jobs. I'll show that additional expense on Figure 9-1.

AEI Builders

Company Profile

AEI Builders is an owner-managed construction company that employs two project manager/estimators, four superintendents, five clerical employees and a field force of 30 to 40 employees.

Only field employee time, including superintendents, is charged to project cost. All other salaries are treated as general overhead. Clerical wages historically run approximately 2 percent of sales. Office facilities, which the company owns, are larger than they need for their current staff.

Construction equipment owned, if fully used, could be costed to jobs at $50,000 per month. The company can bond projects amounting to $3 million in uncompleted work at any one time.

The company's market area will allow 10 percent growth in the upcoming year at current margins. Each project manager/estimator can handle approximately $3 million worth of construction yearly.

The company operates in a projected long-term growth area with demand for its type of construction activity expected to be above average.

The company is assuming inflation in the upcoming year to be zero.

AEI Builders has developed a long-term business plan calling for doubling volume and tripling profits in the next six years. AEI Builders currently gets all their work by bidding.

Figure 9-2 AEI Builders company profile

Our Sample Company

If you're like me, you learn quickest from examples. That's why I include plenty of examples in this book. In the following chapters, the company I'll use in my examples is called AEI Builders. AEI builds single-family homes, apartment projects, and does tenant improvement work in industrial buildings. I'll give you a profile of the company and then we'll use this information to work through some examples that you can apply to your own company's operation.

Figure 9-2 is the AEI Builders' company profile. We'll use the information in that illustration to develop a long-term business plan, and then a yearly business plan, or budget.

AEI Builders Income Statement		
Gross sales		$4,874,000
Construction costs		
Labor	940,000	
Material	1,023,000	
Equipment	362,000	
Subcontracts	1,912,000	
Total costs		-4,237,000
Gross profit from construction		637,000
Other income: Interest from CD's		34,000
Adjusted gross income		$671,000
Overhead		
Management salary	80,000	
Estimator/project manager salaries	110,000	
Clerical salaries	96,000	
Office expense	48,000	
Legal, audit, miscellaneous	48,000	
Equipment not costed to jobs as expected	114,000	
Total overhead costs		$496,000
Profit before taxes		$175,000

Notes:

1. All labor costs includes fringe benefits and taxes.
2. Equipment overhead cost represents total cost less charged to jobs (unused time).

Figure 9-3 Income statement

The Annual Budget

The annual budget begins with anticipated revenue for the coming year. Take a guess at gross sales for the coming year. Then estimate your hard costs like labor, materials and subcontract costs. What's left is your normal margin. Subtract your expected overhead like office expense and office salaries from that margin. The remainder is your before-tax profit.

The best way to illustrate the budgeting process is to have you do it. Keep in mind that a budget should be ambitious, but also realistic. It's fine to dream, but not when you're writing a business plan. The figures in your budget should be numbers you can expect to see at the end of next year if you follow an aggressive plan. Don't set a financial budget that says, "No matter what happens, we ought to be able to get this much work." That's not a good budget because it doesn't set a reasonable yet challenging goal.

AEI Builders

Balance Sheet

Current assets		
Cash	$112,000	
Accounts receivable	923,000	
Prepaid expenses	11,000	
Total current assets		$1,046,000
Fixed assets		
Facilities	250,000	
Equipment	600,000	
Total fixed assets		850,000
Total assets		$1,896,000
Liabilites		
Current - Accounts payable	738,000	
Long-term - Equipment financing	450,000	
Total liabilites		$1,188,000
Net worth		708,000
Total liabilities and net worth		$1,896,000

Notes:

1. Equipment carried at real market value.
2. Office and office equipment owned and paid for.

Figure 9-4 Balance sheet

There's a problem with budgeting in the construction industry. It isn't the task. Anyone who can add a column of figures can draw up a budget. The problem is the industry itself. Most construction companies don't budget overhead and then compare actual revenue and expense against budgeted revenue and expense each month. Companies in most other industries do that. Construction companies tend to monitor *results* much better than *costs*. We keep track of jobs and contracts. We know how much our payroll is each month. We hope we know what our costs are. But we usually don't compare predicted costs with actual costs each month or each quarter.

The automotive industry budgets for a quarter, then divides the quarter into tracking periods of 10 days each. They keep track of units sold, gross profit, and operating costs. When something starts running over budget, they know it immediately.

How Your Budget Helps

Suppose you've budgeted $4,000 per month for advertising. At the end of the first quarter, you've spent $15,000. You know that if you continue to spend money for advertising at that rate, by the end of the year, you'll be $12,000 over budget for advertising. If advertising represents 5 percent of your gross income, you'd have to bring in an additional $240,000 in receipts to cover that overage. You'd certainly want to know about that in April, not discover it in November. That's the advantage of running your business with a budget. You can compare actual results to predicted results every month. If something goes wrong, you have the rest of the year to correct it.

Looking only at sales can also be a mistake. Suppose last year you billed $750,000. This year you expect sales to reach one million dollars. At the end of December your actual billings were $997,000. You'd probably be fairly pleased with yourself. But at the end of January, your accountant explains that you made 2 percent less last year than you did the year before. "What happened? How can that be?" you say. "What's wrong with our bookkeeping system?"

Keep an eye on your figures

The answer is that there's nothing wrong with your bookkeeping system. The numbers are all there. You just didn't look at the right figures. The bottom line is more important than gross sales. Watch sales, of course. That's the easy part. Keep track of overhead expenses too. But it's your profit margin that's the key to survival, and to get that you must subtract expenses from sales.

Put it on paper

Even if you don't have a written business plan, put your annual budget on paper. Then keep track of whether things are going as you planned. If they're not, do something to get back on track. That's management: You set up a business plan, track progress against that plan, and make corrections when they're needed.

Contractors today have to do a pretty good job of controlling individual projects. In the 1970s, that wasn't always true. Many of even the largest building firms were run by seat-of-their-pants managers. They didn't know whether the job made or lost money until all the numbers were in from the bookkeeping department.

Today, most companies must control job costs more carefully. You estimate each part of the project and track costs for all of the tasks — this much for electrical, this much for suspended ceilings, etc. Then check expenses against your estimate as you go along. That way, when job costs exceed your estimate, you can pinpoint the job area and adjust your estimating system accordingly for particular activities.

AEI Builders		
Budget for the year _______		
Gross sales		________
Construction costs		________
Construction profit		________
Overhead		
Management salary	________	
Estimator/project manager salaries	________	
Clerical salaries	________	
Office expense	________	
Legal, audit, misc.	________	
Equipment	________	
Total overhead		________
Profit (pre-tax)		________

Figure 9-5 Annual budget exercise

If I'm your project manager, it's my job to get the project done on time and within budget. I know how much it's supposed to cost, and how much we sold it for. I order the materials, hire the subs, make sure they perform, and schedule everyone on the job to see that they finish on time.

The contractor (or estimator) has the same responsibility for the whole company. The first step is to put your annual overhead budget down on paper.

The Information You'll Need

Figure 9-2 is a profile for AEI Builders. It sets a company goal of doubling sales and tripling profits in the next six years. Figure 9-3 is a company income statement for the last year. Figure 9-4 is a balance sheet for year end. These financial statements and the company profile should help you make some reasonable assumptions about AEI Builders.

Exercise #1

Use Figure 9-5 to prepare a budget for AEI Builders for the coming year. This budget should reflect what you know about the company from Figures 9-2, 9-3 and 9-4. Don't look at my answers until you complete the budget for AEI Builders. Remember that good budget figures are both realistic and ambitious.

AEI Builders **Budget for the year ______**		
Gross sales		6,000,000
Construction costs		5,200,000
Construction profit		800,000
Overhead		
Management salary	80,000	
Estimator/project manager salaries	120,000	
Clerical salaries	120,000	
Office expense	50,000	
Legal, audit, misc.	50,000	
Equipment	0	
Total overhead		420,000
Profit (pre-tax)		380,000

Figure 9-6 Annual budget answers

When you've filled in all the blank lines on Figure 9-5, compare them to my answers in Figure 9-6. Then read my explanation for the numbers in Figure 9-6.

Remember that there's no single right answer to this exercise. It's not the same as estimating how many 6 x 8 x 16 concrete blocks there are in a wall. If you know the length and height of the wall and that there are 1.125 blocks per square foot, your answer will be exactly the same as mine. On Figure 9-5 a range of answers could be right for any line.

My Budget for AEI Builders

In the section that follows each exercise, I'll explain where my answers came from. In some cases, the "correct" answer will be a range, not an exact number, so consider your responses correct if they're within my range. But remember too, that any budget for your company has to come from your own experience. The purpose of these exercises is to demonstrate the budgeting process, not to find exact solutions.

The answers in Figure 9-6 are based on the limited information we have about AEI Builders. When you set up a budget for your company, you'll have much more information to work with. The correct range of answers should be much narrower than in this exercise.

Gross Sales

This is the number where you have the most discretion — the number with the widest range of correct answers. How much work does AEI Builders expect to sell next year? What's a realistic figure that still challenges the company to reach a specific goal?

Your starting point is last year's income (profit and loss) statement. AEI Builders did a little less than five million dollars last year — $4.874 million, to be exact.

Here's another important fact taken from Figure 9-3, the company profile. In the last paragraph, you see that AEI hopes to double their volume and triple their profits over the next six years.

Volume Is Easy to Control

As the owner of a construction company, you can and should control business volume. You can't control everything, but you can control volume. Suppose you decide to cut back on volume, doing only half as much work next year as you did last year. Is that possible? Of course. It's not only possible, it's also very easy. You just don't bid and sell as many jobs. You stop taking contracts at a certain point.

Now suppose you decide to double business volume next year. Can you do that? Probably. You'll also probably double the headaches and the risk. And, of course, you'll have to get very, very competitive with your bidding. But most contractors could probably do it when the market for construction services is strong.

One way to increase sales - A service company, like a law firm or a dental practice, could possibly cut fees and prices by one half and more than double business volume in a year. A contractor couldn't possibly do the same thing and still make money. Not many service companies could either. But is there any doubt that your accountant would get a lot of new business by cutting his hourly fee by one-half? Absolutely not. He'd probably double his client base.

If you do the same in your construction company, you'll get more work than you can handle. But what will happen to your business? You'll lose your shirt unless you've been getting a huge markup in the past. You can buy volume in the construction industry. But not enough to cut selling prices by one-half, or even one-quarter.

Our point here is that most contractors in most business climates can influence business volume by adjusting prices.

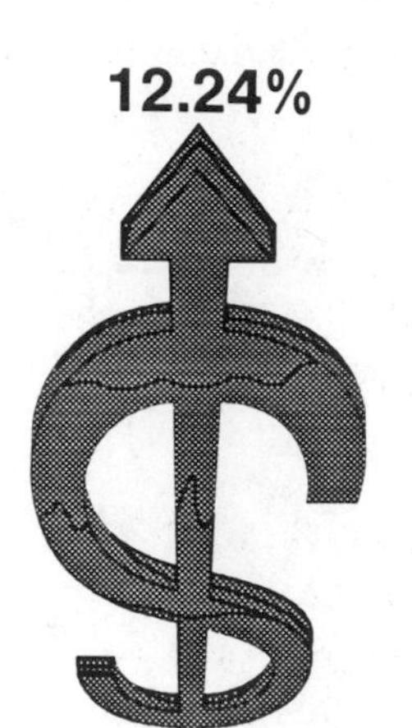

Doubling volume in six years is reasonable - AEI Builders doesn't plan to double its business in the coming year. That's not realistic. It would require pricing that might ruin the company. A more realistic goal is to double volume in six years. If you divide a 100 percent increase by six years, that's a 16.67 percent increase per year. In fact, it's less than that because the increase compounds itself. You'd really only have to increase volume by 12.24 percent each year to be doing twice as much at the end of six years.

Since AEI Builders did $4.874 million last year, a goal of nearly $5.5 million (an increase of about $600,000) would be reasonable for next year, based on the long-term plan. Here's how it works:

✏ 4,874,000.00 x 0.1224 = 596,577.60

✏ 4,874,000.00 + 596,577.60 = 5,470,577.60

What Will the Market Support?

The company profile for AEI Builders (Figure 9-2) stated that the market should support 10 percent growth in the upcoming year at current margins. If you disregard the long-term growth plan, and just use that information, projected gross for the coming year would be about $5.36 million.

✏ 4,874,000 + 487,400 = 5,361,400

Maximum Projected Sales

These calculations produce what I consider to be *conservative* expectations for growth. But AEI should probably look at a more challenging projection for the coming year. That's because there will be good years and bad years during any five-year period.

Most contractors would be happy with growth of 10 percent in a year. But AEI's long-term plan calls for an average yearly growth of a little over 12 percent. So their goal for the coming year should be *more* than that to make up for the slow years. That's why we want to aim a little on the high side in the next year.

Figure 9-2 identified two restrictions on growth at AEI Builders. One was bonding restrictions. The other was personnel. Let's explore each of these.

Bonding limits - The company profile says that AEI Builders is bondable for a maximum of $3 million of work in progress. Most construction companies can't sell more than twice their bonding limit in any year. If all AEI work requires a performance bond, they're limited to about $6 million in annual sales. That's not an absolute maximum, of course, but it's a good rule of thumb. If they try to bid on very much more work than that, their bonding company will probably refuse to issue a bid bond.

Company staff limits - You know AEI Builders has two full-time project manager/estimators who are capable of handling about $3 million worth of work each year. Once volume grows beyond that, AEI will need another project manager, or at least an assistant, to assume part of the work now handled by the existing manager/estimators.

Because of bonding and staff limitations, my budget calls for $6 million in annual sales. I'm going to use that figure in the rest of this discussion. If your projected business volume for the coming year was between $5.3 and $6 million, I'd say you interpreted the information correctly and came up with a reasonable answer. Six million also makes a good goal. Realize, though, that just setting a goal isn't enough to make it happen. The company has to make some changes. We'll come to those later.

Forecasting Construction Costs

How much is it going to cost AEI to build what they'll sell for $6 million? The place to start these calculations is also with Figure 9-3, last year's profit and loss statement. Total construction costs last year were $4,237,000, or about 87 percent of total revenue.

✏ 4,237,000 ÷ 4,874,000 = 0.8693

That means overhead and profit averaged 13 percent of gross revenues last year. That doesn't mean AEI bid their jobs at cost plus 13 percent. There are cost overruns and unplanned expenses to consider, but what they actually ended up with was 13 percent gross profit from construction.

There are two ways to increase sales for any business:

Two ways to increase sales for any business

1) *Spend more*. You can spend more money on promotion or more on doing quality work or more on supervision and management or more on customer satisfaction or more on anything that will build your reputation as a quality builder that the public can rely on.

2) *Charge less*. You can cut the profit margin, reduce expenses and control costs so your bids are highly competitive. Anyone who's price shopping is sure to be interested in your bids.

Either of those strategies will increase volume eventually. Charging less tends to get quicker results and spending more tends to help build a reputation over time. But don't try to do both at once. You can't do top quality work at a bargain basement price. Nobody can. The two are almost mutually exclusive. So how do you get more work?

You get more work by making trade-offs between price and quality as the market permits and as your customer base requires.

My budgeted gross sales figure of $6 million next year is a 23 percent increase over the year before. With growth like that, I'll probably have to cut a few corners and watch costs carefully just to maintain my present profit margin.

If you projected sales of less than $5.5 million for next year, I'd say you're planning to mark time rather than grow with the expanding construction market. If so, it's reasonable to either increase your profit margin by a percent or two or to place more emphasis on quality construction in an effort to build customer satisfaction.

In my budget, I'm going to stay with the 87 percent figure for construction costs. That rounds to about $5.2 million, and a gross profit of about $800,000.

Higher volume means lower gross margin - Big increases in volume, whether planned or not, usually reduce the profit margin. The first reason for this is obvious. Every dollar cut from bid prices also cuts a dollar out of profit margin. The second reason is less apparent. When your company is unusually busy, you have to hire new people and overwork others. The usual result is a sacrifice in control over production costs. That reduces profit margin as well.

When your workload isn't heavy, you have time to pin down the best prices from material suppliers and subcontractors. When you're swamped, you can't always do that. You take the first bid if the price seems reasonable. You place orders without negotiating. You tolerate poor or marginal performance from some employees because it's too hard to find and train a replacement. And in each case, you'll probably pay a few bucks more for what you buy.

Forecasting Overhead

Your margin on a job isn't all profit, of course. Overhead expenses are real costs, just like concrete and lumber. In our example budget, we've broken overhead down into six broad categories for the sake of simplicity. In practice, your books might show as many as 30 or more. But all 30 will fit into some category in our summary. Here are the broad overhead categories.

Management salary - The structure of AEI Builders is typical of the construction industry. That is, it's run by an owner/manager. That's different from most other industries today. Most companies, especially large ones, are run by hired managers rather than stockholder/owners.

Only two other industries, agriculture and food service, have remained largely owner-operated. And even those are changing. Twenty years ago, most farming was done by family-owned businesses. Now, most farms are still operated by the owner. But corporate-owned farms gross well over half the sales dollars from farming.

"The vast majority of all building construction and remodeling companies today are still owned and managed by an individual."

In the restaurant trade, both high- and low-end businesses were individually owned until recently. Now, most of those are run by corporate-owned chains. By contrast, nearly all major manufacturers are owned by stockholders who never set foot on the premises. Steel, automobile, and transportation giants were never family-owned businesses. In the airline industry, for instance, the ten majors do 90 percent of the total business. Not so in construction. There will probably never be a company the size of General Motors or Ford that grabs 30 or 40 percent of the construction market. All of us in this business can be glad of that.

The vast majority of all building construction and remodeling companies today are still owned and managed by an individual. Even many of the major home builders aren't publicly-held companies. They're family- or privately-owned corporations or partnerships. The few that are publicly owned aren't really contractors at all. They're land development firms that invest stockholder money in land and then contract out the actual construction.

Notice that the first line under overhead is management salary. In most industries, management salaries are based on performance and market conditions. The construction industry is different. In construction, management salary is usually the owner's draw. It also includes fringe benefits (such as a health plan) and other costs that relate directly to the manager's activities, such as a company-owned vehicle the manager uses to travel between jobs.

Most of the money in the "Management salary" category is the owner's wage. And it's not usually determined by performance or a percentage of gross receipts. It's usually based on:

1) How much the owner needs to meet family obligations.

2) A judgment (usually based on an accountant's advice) of the salary that will leave adequate working capital in the business but still minimize taxes for both the company and the owner or owners.

***How much does the owner really make?* -** What did the manager in our sample company really earn last year? If you say $80,000, you've overlooked something. The manager actually earned almost three times that much. There's the $80,000 he drew out in salary, but there's also another $175,000 in pre-tax profit he left in the business. He could have paid himself $200,000 if he had wanted to. But he and his accountant decided $80,000 was the best figure after considering the tax alternatives and the company's need for cash. AEI may have plans (such as buying a new truck or tractor) that require some extra cash during the coming year.

For this exercise, I'm going to leave management salary at $80,000. That's because AEI's projection is for a zero inflation rate for the coming year. Of course, that's seldom realistic. If you're living on $80,000 this year, you might need $85,000 in the coming year to enjoy the same standard of living.

***The senior estimator's salary* -** AEI Builders is a small company. A construction company isn't considered large until gross receipts pass $30 million a year. At that point, they're probably managed by a team of vice presidents who share the management salary. But most small- and medium-sized companies are run by one person. And right below that person there's usually one or more estimator/project managers. The senior estimator is the second most important person in the company. It's his or her responsibility to get the work in the first place, and then to oversee the jobs on a daily basis. The senior estimator does the estimating, negotiates with the subcontractors and manages the superintendents.

We know that AEI Builders has two estimator/managers. Their combined salary was $110,000. That figure includes wages, benefits, taxes and insurance (labor burden), and the use of company-owned vehicles. It probably also included some incentive pay. Most companies pay their senior estimator/project managers a bonus based on how well they do and how successful the company is during the year.

For next year, I'm going to increase the budget for this category to $120,000. That's because we're going after more work. That costs money. It costs more to bid more jobs. Either you have to bid more jobs or you have to bid larger jobs. If bonuses are tied to gross receipts, bonus pay will go up. AEI managers won't be very happy if they don't get some increase in their regular salary.

My choice of $120,000 is arbitrary. I can't justify anything less than last year since I expect a significant increase in volume of work. On the other hand, I shouldn't increase estimator/manager compensation by any more than the 23 percent increase in sales. Consider your answer correct if it's between $110,000 and $132,000.

Clerical salaries - Here's a case where there's really only one choice. The company profile says that clerical wages historically run 2 percent of sales. That covers gross pay, including benefits and burden for the bookkeeper, secretary, and other office staff.

The company may need to hire another purchasing clerk or bookkeeper to handle the additional volume, and the current employees will expect raises. So I'll use $120,000 for clerical salaries.

Office expense - You didn't have much to go on here. We know that AEI Builders owns its office. In fact, AEI has more space than it currently needs. So there's no increase in cost here. But expect a small increase in telephone and other utility expenses. You could leave office expense unchanged or increase it by not more than 10 percent. An answer between $48,000 and $53,000 is justified. I rounded last year's figure up to an even $50,000.

Legal, audit, miscellaneous - We didn't have much information here, either. You could have left this figure unchanged or built in a small increase. I rounded this category to $50,000.

Equipment not charged to jobs - There's only one right answer to this. The answer should be zero. Notice, this isn't office equipment — it's heavy equipment used in construction. All those costs should be charged off to jobs, with none charged to overhead as unused time.

Questions to ask yourself

Your first step is to anticipate what it will cost to own each piece of equipment for the coming year. Suppose you're an earthmoving contractor and you own a D-6 dozer that's a year and a half old. Here's what you have to ask at the beginning of the year:

- How much interest will I pay?
- How much will I charge to depreciation?
- What taxes and fees do I have to pay?
- What do I expect maintenance to cost?

Suppose your total is $50,000. Now you need to estimate how much you'll use that equipment during the year. For instance, figure you'll operate the dozer about half the time, or for 1200 hours. The rest of the time it will sit idle in the yard waiting for another job to begin. Divide $50,000 by 1200, and you get an hourly billing charge of around $40 ($41.67 to be exact).

Now, you instruct your estimators to use $40 per hour as the internal billing figure for that piece of equipment. (You're gambling that you'll actually use the equipment a little over 1200 hours.) You also tell your project manager that his jobs will be charged $40 for every hour that piece of equipment is in use. At the end of the year, if your estimate is accurate, you'll break even on that machine. All costs of owning that dozer (plus your normal markup) get passed on to jobs where the dozer was used.

AEI Builders made a mistake on equipment last year. Their equipment ownership cost was $476,000 and only $362,000 of that was charged off to jobs. The rest turned up on the income statement as overhead.

If you have the same problem, here are three ways to solve it.

1) If you underestimated the cost of ownership, you'll have to increase your billing rate. Suppose you were right about using the equipment 1200 hours during the year. But interest, depreciation taxes and maintenance actually cost $56,400, not the $50,000 you anticipated. The result? You have to increase the billing rate from $40 to $47 per hour.

2) If you have equipment sitting idle, change the type of projects you're bidding. Go after jobs that can use the equipment you have. Dirt contractors with a yard full of idle earthmoving equipment don't make a dime hanging drywall.

3) If you can't come up with enough projects to make reasonable use of your equipment, rent it out or sell it.

Because we're just working on the preliminary budget now, we won't decide yet which of the three choices is the best. But this is the time to make a note: Something has to change next year so there's no equipment backcharge at year-end.

The Preliminary Budget Bottom Line

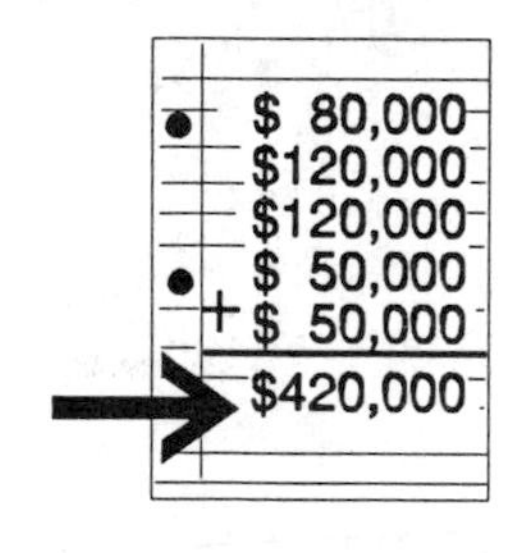

My preliminary budget shows total overhead of $420,000. That comes out of the $800,000 margin (gross profit) and leaves a net pre-tax income of $380,000.

Now that we have the budget, what do we do with it? What's the main reason for preparing an annual budget? Easy! It's a guide for reaching long-term goals. The budget must meet those goals. To see if it does, ask these questions:

Ask yourself these questions

1) Is our projected gross sales total in line with our long-term plan?

2) Does our expected profit fit our long-term plan?

We've already discussed gross sales. Our 23 percent projected increase is well over the 12.24 percent we need to double gross volume in six years.

Focus on Profit

Don't focus just on sales. Profit is even more important. AEI Builders plans to triple its profit in six years. That requires an increase of about 20 percent each year. Again, these increases are compounded. The annual increase can be much less than the 33⅓ percent required if you simply divide a 200 percent increase by six years. If you increase profit by 20 percent each year, six years from now your profit will be just about three times what it is today.

At first glance, it seems that AEI Builders expects to more than double its profit over last year. And that's true. But it's also misleading. Back up and try to guess what AEI planned to make (not what they actually made) in the previous year.

Remember that $114,000 overhead expense for equipment? That's money they didn't expect to absorb last year. They expected those costs to be paid by jobs. If the year had gone according to plan, actual profit for the year would have been $289,000. Next year's projected profit of $380,000 is only 31.5 percent more than they projected for last year. It's also more than enough to meet their long-term goal.

Notice that the budget is based on last year's *plan*, not last year's *results*. We don't have enough information to be sure that everything besides equipment expenses went according to budget. But we'll assume it did, and AEI Builders planned to make $289,000 last year. You'll never meet your long-term business plan goal if you base next year's results on last year's mistakes. That defeats the purpose of the plan. Stick to the plan if you want to reach the goal.

Is Our Long-term Goal Realistic?

On paper this looks like a reasonable goal. To make sure, let's compare it to some industry averages. Several national surveys of construction contractors have reported that general contractors average about 3 percent pre-tax profit. That figure hasn't changed much over the last 40 years, through good times and bad. And it includes companies that have lost money or even gone belly-up during the year. Half of all construction firms fail within any five-year period. So if you're surviving, you're probably doing better than average.

Remember too that the manager's salary isn't profit. Some very small contractors consider their draw and the company's profit to be the same thing. But true profit is what's left after you pay yourself a reasonable wage — the same amount you would have to pay someone else to do the work.

"Half of all construction firms fail within any five-year period."

If you're one of the lucky few who made a 10 percent profit on sales for the year, pat yourself on the back. You're thriving in an industry where low profit margins are the rule rather than the exception.

Recommended changes for AEI Builders
In the space above, list three or more annual business plan recommendations for changes in staff, market direction, facility improvements or operating efficiency.

Figure 9-7 Business plan recommendations

The owner of AEI Builders should be very pleased if results for the year match the budget. Profit represents 6.33 percent of gross sales, well above the industry average. And, it meets the business plan goal of tripling profits in six years because it's more than 20 percent higher than planned profit for the previous year.

The budget projects good growth and improved profits. But I'll be the first to admit that writing these numbers down doesn't make them happen. The owner and chief estimator have to make them happen. Let the business bump along as usual and it'll never happen.

Make Some Changes

If AEI Builders doesn't change anything, gross income for next year will be about $5.3 million. That's a reasonable expectation given market conditions in their location. But they'll need to make changes to raise receipts to $6 million.

Now, go back and look at Figures 9-2 through 9-4, the company profile, the income statement, and the balance sheet. On Figure 9-7, write down the recommendations you'd make to help AEI Builders reach their sales and profit goals for the coming year. Concentrate on the long-term goal. That's more important than profit or loss in any particular year.

I see three or four fairly obvious changes that AEI Builders must make in the coming year. Write down your changes before reading any further.

AEI's Changes

You might have included something like, "Reduce job costs through better contract negotiation." That's possible. But that's not a key problem for AEI, and it's not a one-shot change they can make immediately. Better contract negotiation is something all construction companies should work on all the time. So I didn't include it in my list. Let's look elsewhere for changes.

My suggestions for AEI are listed in Figure 9-8. I've listed suggestions for changes in six categories. Two of them need further explanation: training and bonding.

Suggestion: Begin training - First, I note that AEI Builders is ambitious. They plan to double volume. So they'll need twice as many qualified supervisors. Otherwise, lack of trained people will stifle growth. They'll also need at least one more estimator/project manager.

Two years from now they project sales well in excess of $6 million. Can the present staff handle that extra business? Don't count on it. Maybe it's just as important to train supervisors as it is to manage the equipment better.

You know from your own experience that construction companies can't hire the qualified supervisors they need on a moment's notice. Instead, we have to train people. That takes time. The more complex the job, the higher our standards, the longer that training will take.

Recommended changes for AEI Builders
Promote or hire and train additional supervisors
Eliminate the equipment backcharge by selling some equipment, raise the equipment billing rate, or solicit jobs that make better use of company-owned equipment.
Begin training another estimator/project manager to handle the increase in business that the long-term plan precribes for future years.
Plan a more aggressive marketing strategy. We need to beef up our sales efforts to bring in the additional work we need to reach our budget predictions.
Increase bonding to allow for more than $3 million in business at any given time.
Hire one or more full- or part-time employees to handle the increase in bookkeeping and other clerical activities.

Figure 9-8 Business plan recommendations (answers)

Since training takes time, build it into your plan this year if you expect results next year. Either begin training people in the company now, or hire them from outside. If you select current staff members, then you'll have to hire or promote others to fill the vacated positions. Plan to do that during the coming year. Don't wait until your present staff is overwhelmed with work.

Later, we'll discuss some specific ways to bring more work into the company. For now, just realize that the additional revenue won't happen unless you make it happen. And when it does happen, it brings extra headaches as well.

Suggestion: Increase bonding capacity - AEI Builders expects to do $6 million worth of business next year. The company profile says they're bonding line of credit is $3 million. Is that enough? Only if they do $3 million in the first half of the year, finish those jobs, then do the other $3 million in the last half. But what if they have to start some of those jobs before they finish earlier jobs. What if there's $4 million or more of work in progress? AEI Builders needs extra bonding capacity to cover that volume. How do you increase bonding capacity? By increasing the net worth of the company. Net worth comes from after-tax profit. So increases in profit (if left in the business) will help increase business volume.

Look over the other suggestions I offer in Figure 9-8. See if you don't agree. Compare my suggestions with your business plan recommendations in Figure 9-7. How are my suggestions different? Which changes do you prefer? And, most important, what changes are needed in your company?

Failing to Plan Is Planning to Fail

That's the simple purpose of this chapter: To help you recognize needed changes, not just at AEI Builders, but in your company. Once you recognize what changes you have to make, making them is both more likely and much easier. If you don't recognize the need for changes, if you therefore don't make those changes, you're going to continue doing no better than you're doing now. If you're satisfied with that, fine. I'll never convince you otherwise. But if you want something different, the tools I've provided here are the tools you need to make changes for the better in your company.

Chapter 10
Estimating Overhead & Profit

Overhead and profit are usually the last items you add to your bid, almost like an afterthought you're embarrassed to mention. Of course, they're not. Company overhead is your cost of doing business. Profit is your reason for doing business. Forget either and you won't be in business long.

Sometimes you'll hear the terms gross profit and net profit. The profit we're talking about in this chapter is net profit. That's what's left after you pay all the bills, including your overhead costs. Net profit is what the company pays taxes on. It's also the return on money invested in the business.

The term gross profit is used more often by accountants than contractors. By gross profit, they mean what's available to pay bills after job expenses are paid. Used that way, gross profit includes both company overhead and net profit. Another word for gross profit is *margin*.

The purpose of this chapter is to answer a simple question: What percentage should you add to your estimates to cover company overhead expense and provide a reasonable profit?

It isn't always easy to answer that question. But if you've tracked actual overhead for the last year or two and have an overhead budget for the coming year (as recommended in Chapter 9), you've got half the answer already. The other half is profit. You know the profit you'd like to make. That's in your long-term business plan. But what's the best profit you can actually make this year? That's the question this chapter will answer.

Your Normal Markup

You don't have to guess about the best profit margin to include in your estimates. There's a very good way to determine exactly how much you should charge for profit. If you're charging less, you're losing part of the money you could make in this business. If you're charging more, you may be losing some profitable jobs. If you don't believe that it's possible to calculate the "right" profit percentage for an estimate, keep reading. By the end of this chapter, you'll be convinced.

Most construction companies figure overhead and profit as a percentage of job cost. Some companies use another method — for instance, they add a fixed fee to the bid. But most bids include overhead and profit (collectively called *markup*) as a percentage of cost.

Two ways to apply markup as a percentage

There are two ways to apply markup as a percentage:

1) Add the same markup percentage to each of your bids, or,

2) Vary your markup with the type of job or the bid item (material, labor, equipment or subcontract cost).

Your pricing method depends on the type of work you do. If you bid the same type of work nearly all the time, you'll probably feel comfortable using the same markup percentage on nearly all jobs. For example, if you do only kitchen and bath jobs in a certain price range, all of your jobs may be about 20 percent labor, 30 percent materials, 40 percent subcontract work, and 10 percent equipment. If that's the case, applying different markups to different parts of the job wouldn't make much difference. You can average the markups and use a single percentage. Here's an example:

I'm bidding on a single-family custom home. My total cost is $120,000. This house is typical of the work I do. My cost breakdown looks like this:

Category	Percent of Total	Cost	Percent Markup	Sale Price
Labor	20%	24,000	22%	29,280
Materials	30%	36,000	15%	41,400
Subcontracts	40%	48,000	11%	53,280
Equipment	10%	12,000	20%	14,400
Total	100%	120,000	15.3%	138,360

If my business plan requires a 15 percent markup, my task is easy. My bid for this job is $138,000. But if I do other types of work that are mostly subcontract or involve a lot of equipment, I'll probably apply a separate markup to each of the four cost categories (labor, material, equipment and subcontract).

Most contractors use a higher markup for labor and equipment cost than for subcontract or material cost. That's because there's more risk on labor and equipment cost estimates, and they're related to each other. You can be pretty sure your estimates for subcontract and material costs are accurate. But there's more danger of an underestimate on labor and equipment costs. A mistake, an accident, or a delay can put those costs out of line in a hurry.

Here's another example:

Suppose the job I'm bidding now includes a heavy percentage of labor costs. I'll use my own crews instead of subbing parts of the job out. If I use a flat 15 percent markup, here's how the job will come out:

Category	Percent of Total	Cost	Percent Markup	Sale Price
Labor	60%	72,000	15%	82,800
Materials	30%	36,000	15%	41,400
Subcontracts	0%	0	15%	0
Equipment	10%	12,000	15%	13,800
Total	100%	120,000	15%	138,000

But if I vary the markup for each of the cost categories in proportion to risk, it might come out this way:

Category	Percent of Total	Cost	Percent Markup	Sale Price
Labor	60%	72,000	22%	87,840
Materials	30%	36,000	15%	41,400
Subcontracts	0%	0	11%	0
Equipment	10%	12,000	20%	14,400
Total	100%	120,000	19.7%	143,640

If I get a lot of these "unusual" jobs, my gross profit for the year should be better than expected, provided I vary the markup.

If you handle a wide variety of jobs, I recommend that you develop some written guidelines for applying markup. For instance, you might want to set a limit on markup such as:

Markup is 15 percent of total cost (labor, material, equipment and subcontract) but will never be more than 50 percent, nor less than 25 percent of the labor cost alone.

The Right Markup for Your Estimates

Later in this chapter you'll create a markup vs. volume graph to determine the correct markup for AEI Builders. I expect you'll want to use the same method to decide on the right markup for your company.

The theory behind a markup vs. volume graph is that volume and markup vary inversely. You'll lose more jobs as markup increases and win more jobs as markup falls. We'll use the markup vs. volume graph to create a "what-if" model and determine the best markup-to-volume ratio for your company.

Keep a Bid History

You've heard the old saying, "Those who don't understand history are doomed to repeat it." To avoid repeating your bidding mistakes, keep a written record of your bids, like Figure 10-1. This is a bid history for our imaginary company, AEI Builders. The *Bid price* column shows the selling price for each job they estimated. The *Low (or 2nd)* column shows the low bid for jobs AEI Builders didn't win (or the next lowest bid if AEI got the work). The *% Over* column indicates the percentage difference between AEI Builders' proposal and the next lower or winning bid. The bold lines indicate the bids on which AEI Builders was successful.

Look at the first line, the January 10 job. You see that all costs, including job overhead, came in at $90,000. That cost doesn't include office overhead or estimating time. AEI Builders' bid on the job was $103,500, but the job went for $98,000. That's $5,500, or 5.3 percent less than our bid. AEI would have had to bid that job at less than 10 percent markup to win it.

Now look at the third entry. Cost for the January 14 job was $269,000. AEI Builders bid $317,500, and the next lowest bid was for $334,000. AEI Builders could have bid 5 percent more, and still gotten that job.

It's not always easy to assemble the kind of information you see in Figure 10-1. Every time you submit a bid, you have to find out what the job went for (if you didn't get it) and what the next lower bid was (if you

BID HISTORY						
Date	Type	Location	Cost (000's)	Bid price	Low (or 2nd)	% over
10-Jan	House	Sarasota	90	103,500	98,000	5.3
12-Jan	House	Sarasota	122	140,000	139,000	0.7
14-Jan	**Office**	**Sarasota**	**269**	**317,500**	**334,000**	**-5.2**
28-Jan	Apartment	Tampa	845	998,000	949,000	4.9
5-Feb	House	Sarasota	163	188,000	179,000	4.8
7-Feb	**House**	**Sarasota**	**97**	**109,000**	**112,000**	**-2.8**
14-Feb	Apartment	St. Pete	594	689,000	674,000	2.2
20-Feb	Office	Naples	390	452,000	445,000	1.5
21-Feb	House	Sarasota	74	85,000	79,000	7.1
5-Mar	**Apartment**	**Sarasota**	**1055**	**1,224,000**	**1,249,000**	**-0.4**
14-Mar	House	Sarasota	148	170,000	159,000	6.5
21-Mar	**Apartment**	**Sarasota**	**773**	**894,000**	**899,000**	**-0.6**
28-Mar	Office	Tampa	410	484,000	478,000	1.2
1-Apr	House	Tampa	202	232,000	199,000	14.2
15-Apr	House	Sarasota	91	103,000	102,000	1.0
20-Apr	**House**	**Sarasota**	**110**	**128,000**	**139,000**	**-8.6**
8-May	Office	Sarasota	515	605,000	589,000	2.6
16-May	**Office**	**Sarasota**	**261**	**299,000**	**331,000**	**-10.7**
23-May	House	Sarasota	95	113,000	99,000	12.4
2-Jun	House	Tampa	170	195,000	189,000	3.1
15-Jun	**House**	**Sarasota**	**90**	**105,000**	**106,500**	**-1.4**
10-Jul	**Office**	**Sarasota**	**355**	**420,000**	**435,000**	**-3.6**
24-Jul	Apartment	Sarasota	1520	1,784,000	1,740,000	2.5
15-Aug	House	St. Pete	148	170,000	149,000	12.4
22-Aug	House	Sarasota	130	149,000	139,000	6.7
3-Sep	House	Sarasota	72	83,000	82,500	0.6
15-Sep	Office	St. Pete	1140	1,338,000	1,309,000	2.2
29-Sep	**Office**	**Sarasota**	**1015**	**1,169,000**	**1,249,000**	**-6.8**
10-Oct	Apartment	Sarasota	630	741.000	725,000	2.2
23-Oct	House	Sarasota	68	79,000	75,000	5.1
3-Nov	House	Tampa	162	189,000	159,000	15.9
15-Nov	**House**	**Sarasota**	**145**	**167,000**	**170,000**	**-1.8**
1-Dec	Apartment	Sarasota	910	1,048,000	1,047,000	0.1
12-Dec	Office	Tampa	253	292,000	289,000	1.0
	Totals		13,112	15,283,000		

Note: Tampa and St. Petersburg are 60 miles from AEI Builders.
Naples is 100 miles away.

Figure 10-1 AEI Builders bid history

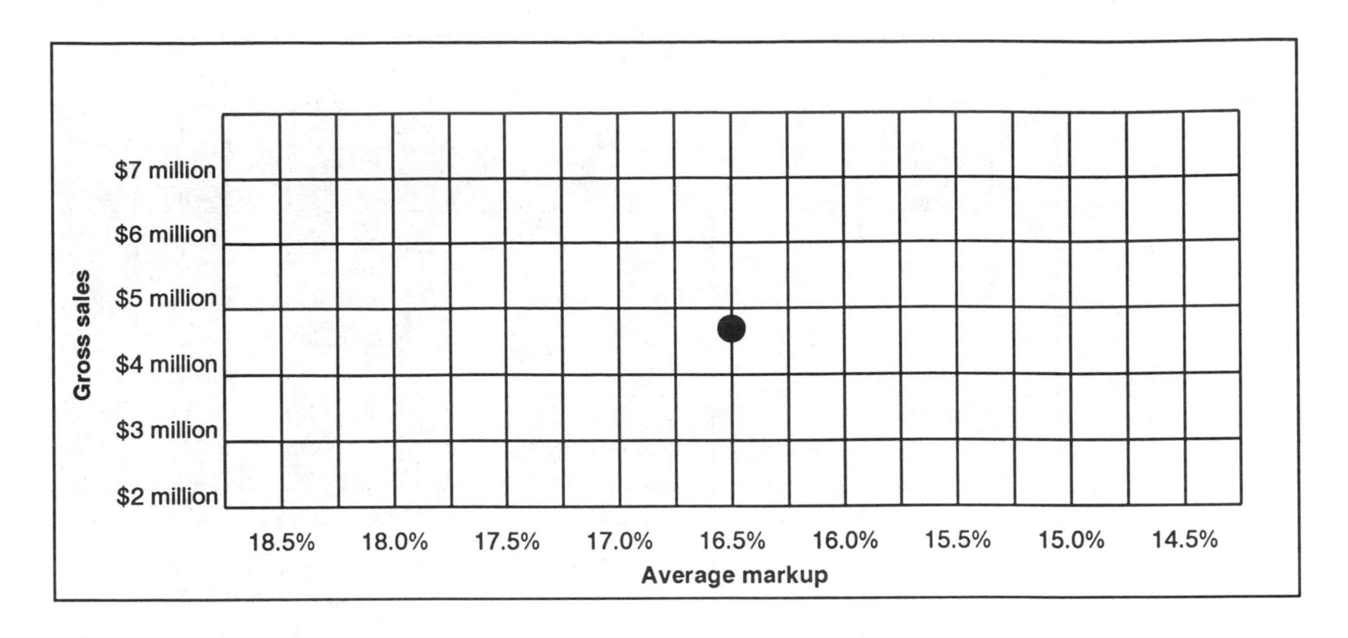

Figure 10-2 Markup vs. volume

ended up doing the work). If you bid on government jobs or other "open bidding" jobs, it's easy to get the information. You just go to the bid opening or ask the owner (or owner's representative). It's a little harder to find out what your competition did in a closed bid situation. But there are ways to collect this information and we'll discuss them in Chapter 12.

For now, I'll assume you've put together a list like the one in Figure 10-1. Here's how to use it to figure out the right markup for your company.

The Markup vs. Volume Graph

First, determine from AEI Builders' bid history what their average markup was for the year. Notice that markup for the year is not necessarily the markup they used on every job. It's the average markup they used for all the jobs they bid during the year.

Here are the totals of the *Cost* and *Bid price* columns for all jobs:

✏ Bid price $15,283,000

✏ Cost $13,112,000

Subtract the cost from the bid price and divide the difference by the cost total. That's the average markup for all jobs:

✏ $15,283,000 - $13,112,000 = $2,171,000

✏ $2,171,000 ÷ $13,112,000 = 0.1655 (or 16.5%)

We've placed a mark on Figure 10-2 where 4.87 million (gross sales for the year) and 16.5 percent (the average markup) meet. You begin there because AEI Builders did $4.87 million worth of work at an average

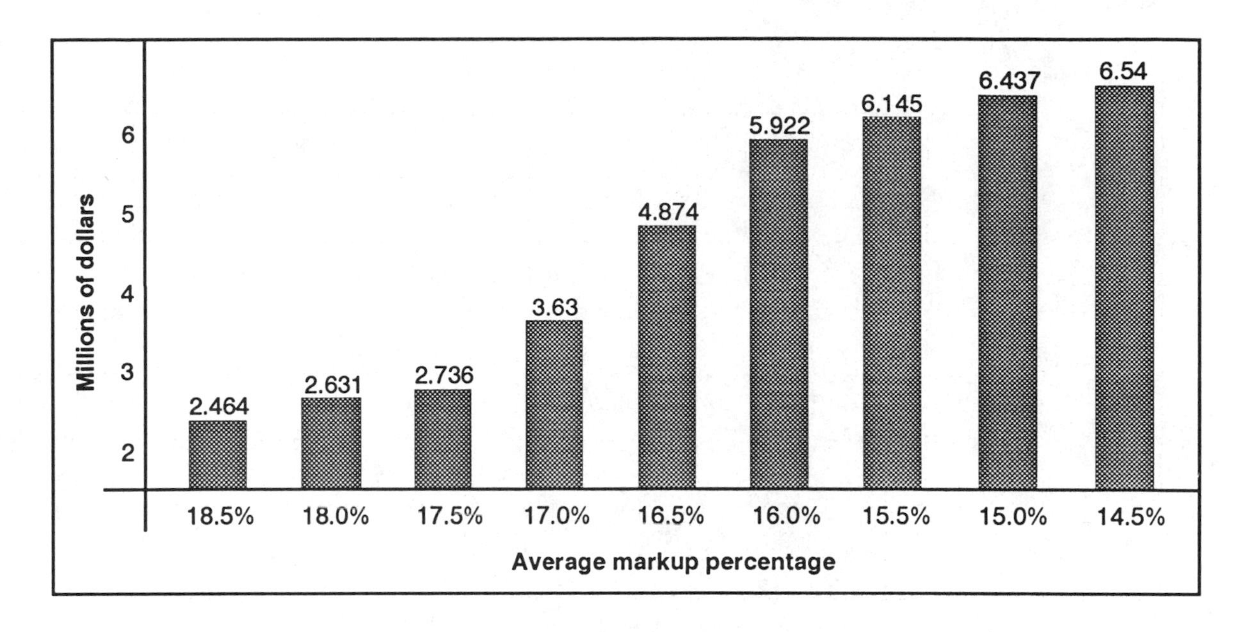

Figure 10-3 Markup vs. volume graph

16.5 percent markup for the year. When you do this for your own company, begin your graph in the middle, with a point where your gross sales total meets your average markup.

Assume markup had been raised to 17 percent. Any jobs that AEI Builders won by less than half a percent would have gone to the competition instead. How much business would AEI have done at 17 percent markup? To find the volume at 17 percent markup, deduct from last year's volume total the jobs that AEI Builders won by less than half a percent.

There's one job in the 17 percent column — the apartment job on March 5. If AEI Builders had bid that job at a half percent more, they would have lost it. So at an average 17 percent markup, total volume for the year would have been $3.63 million. Let's plot these numbers to see if a pattern emerges.

Look at Figure 10-3. Notice that volume is $4.874 million at 16.5 percent markup. That was actual volume last year for AEI Builders. At 17 percent markup, volume would drop to $3.63 million, as Figure 10-3 shows.

If AEI had raised their markup a full percent, volume would have been only $2.736 million. I've plotted that figure above 17.5 percent markup.

At a 1.5 percent increase, AEI Builders would not have won the June 15 job. At 2 percent higher, they would have lost the November 15 job. Their volumes at those percentages would have been $2.631 million and $2.464 million respectively.

You can see that higher markups reduce sales volume. If AEI Builders had increased their average markup by 2 percent, volume would have dropped by almost half.

I've also extended Figure 10-3 in the opposite direction. I looked for jobs AEI Builders *lost* by less than half a percent. I added the value of those jobs to $4.874 million and marked the total in the 16 percent column. I've done the same for markups down to 14.5 percent, working to the right.

At 16 percent, AEI Builders would have picked up the December 1 job, adding $1.048 million to volume for a total of $5.922 million. At 15.5 percent, they would have added a total of $223,000 more for the January 12 and September 3 jobs, for a total of $6.145 million. And at 15 percent they would have picked up another $292,000 for the December 12 job, plus another $103,000 for the April 15 job at 14.5 percent. If they had won all the jobs they lost by less than 2 percent, their total volume for the year would have been $6.540 million.

It should be clear from Figure 10-3 that markup affects volume. If we raise prices, we get less work — if we lower prices we get more.

But Will We Make More Money?

Figure 10-3 shows the relationship between markup and volume, but it doesn't say anything about profit. If I continue lowering the markup, eventually I'll get every job in the state. But I won't make any money.

Raising my markup would have the opposite effect. At 50 percent markup, AEI builders might get one or two little remodeling jobs. With an overhead of half a million dollars, AEI would be bankrupt in short order.

Develop a Profit Curve

The markup vs. volume graph (Figure 10-3) is very useful. But it doesn't tell us much about profit. And profit is more important than volume. Let's create a profit curve for AEI Builders.

Every business has a profit curve, just like every business has a balance sheet and a profit and loss statement. Even if you've never seen your profit curve, you have one. Understanding your profit curve will help you identify the right markup for your estimates.

Compare Figure 10-3 (the markup vs. volume graph) with Figure 10-4 (the profit curve for AEI Builders). You'll notice one thing immediately. Figure 10-3 shows an ascending line. Volume rises as markup declines, probably forever. The profit curve is more bell-shaped. There's a peak where the profit is the greatest. Then profit falls off. Right at the peak is where you want to be.

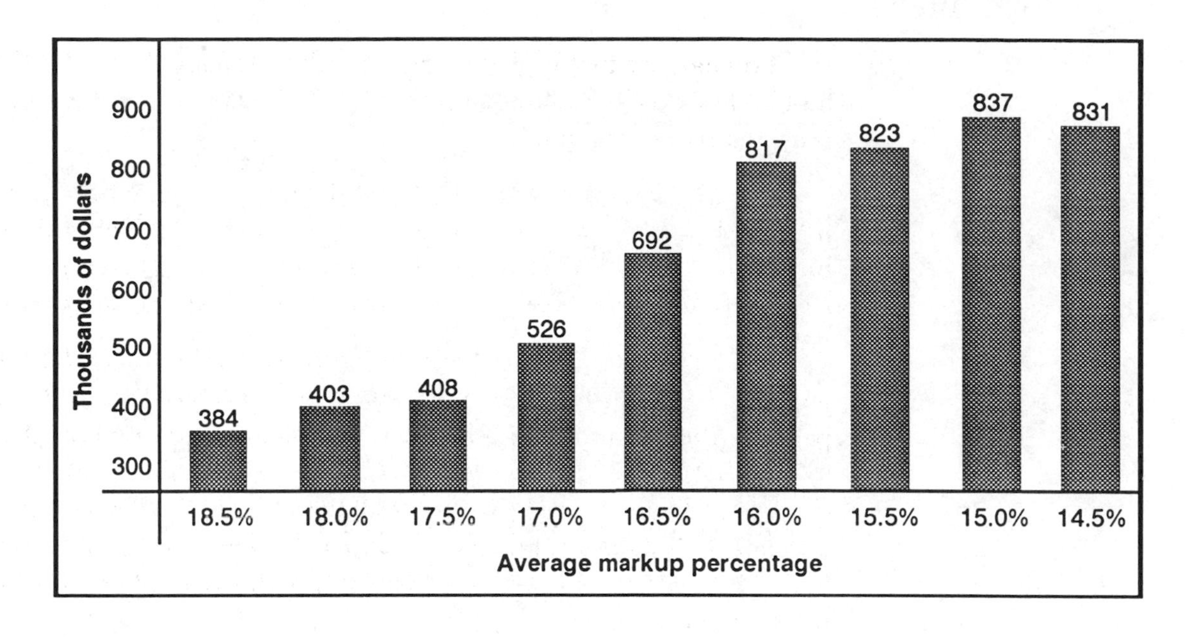

Figure 10-4 The profit curve

The Profit Curve for AEI

Here's how to create a profit curve for AEI. Begin at the 18.5 percent markup point. Suppose AEI Builders had used 18.5 percent markup on all jobs during the previous year. We see from Figure 10-3 that total receipts would have been $2.464 million. What would profit be at 18.5 percent markup and $2.464 volume? It's not $2.464 million times 18.5 percent because markup in percent isn't equal to profit in percent. Here's why:

The percent you add for markup is always more than your profit percent on the job. For example, suppose your markup is 10 percent on job costs of $100,000. The selling price is $110,000. But your profit on the selling price is 9.09 percent, not 10 percent. Here's the calculation:

- Job cost + (10% of job cost) = selling price
- $100,000 + $10,000 = $110,000
- Markup ÷ selling price = profit on the job
- $10,000 ÷ $110,000 = 0.0909 or 9.1%

I've created Figure 10-5 to convert markups in percent to profits in percent. We'll use this figure to calculate the profit in dollars at each volume level for AEI Builders.

The conversion chart shows that 18.5 percent markup yields a 15.6 percent profit. Multiply gross receipts by the profit percentage (15.6) to find the profit in dollars at that volume level. The result is the profit for the year at that volume and markup. For example,

- $2,464,000 x 0.156 = $384,384

Markup		% Profit	Markup		% Profit
10	=	9.1	15.5	=	13.4
10.5	=	9.5	16	=	13.8
11	=	9.9	16.5	=	14.2
11.5	=	10.3	17	=	14.5
12	=	10.7	17.5	=	14.9
12.5	=	11.1	18	=	15.3
13	=	11.5	18.5	=	15.6
13.5	=	11.9	19	=	16.0
14	=	12.3	19.5	=	16.3
14.5	=	12.7	20	=	16.7
15	=	13.0	20.5	=	17.1

Figure 10-5 Markup vs. profit conversion table

That's what the first column in Figure 10-4 shows. Profit would have been a bit over $384,000 for the year.

The next point on Figure 10-3 is 18 percent markup. Multiply $2.631 million by 15.3 percent (from the conversion chart) to find the total profit of $402,543. Plot that amount on the chart and continue plotting the rest of the points the same way. When you're finished, you should have a chart that looks like Figure 10-4.

At the last point on the graph, the curve has passed its peak and is starting down again. Every profit curve does that. The curve goes to a peak somewhere in the center and drops off at both ends with either higher or lower volume. That's because markup eventually goes to zero on the right end and volume eventually drops to zero on the left end.

Where's the Most Profit?

Looking at Figure 10-4, you might conclude that markup should be 15 percent because that's the point where gross profit is greatest. Unfortunately, that's not the whole story. What's missing from Figure 10-4 is our cost of doing business. At lower volume, overhead costs will be somewhat lower. At higher volume, we'll have higher overhead costs. To see the whole picture, we have to add one more line to the graph.

Look at Figure 10-6. The top line is labeled *Gross profit* and is the same as in Figure 10-4. The line labeled *Calculated overhead* is the cost of doing business. It's based on calculations we did in Chapter 9. The space in Figure 10-6 between the *Gross profit* and *Calculated overhead*

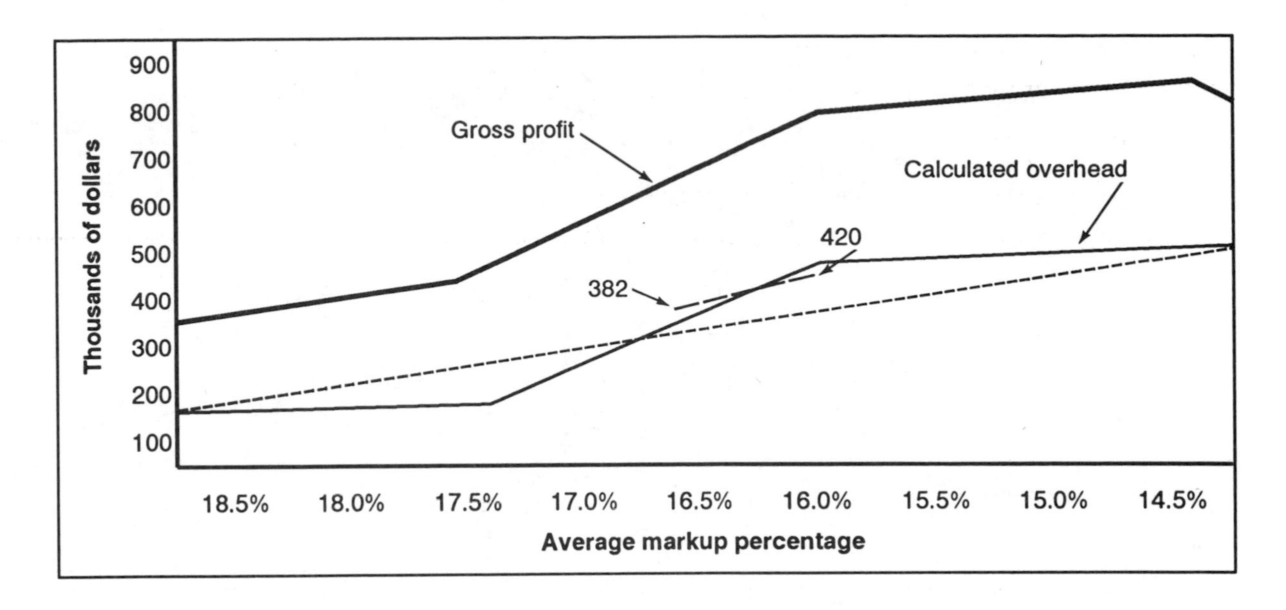

Figure 10-6 Profit vs. overhead

lines shows the net profit after subtracting overhead. Where the two lines are the farthest apart, net profit is the greatest. That's the markup we want to charge.

Now, how did I figure the line *Calculated overhead*? We don't have a chart that shows overhead at every volume level between $2.464 million and $6.54 million. There's no way to be sure what overhead would be at each of those volume levels. But we can make some educated guesses. Here's how:

Estimating Overhead

Remember from our discussion of the annual budget (Figure 9-6) that we projected overhead cost at $420,000 for $6 million in volume. On Figure 10-6 I've made a mark where $420,000 meets 16 percent markup.

On Figure 10-6 I've also plotted the $382,000 overhead figure at 16.5 percent markup. These figures come from the profit and loss statement, less the equipment backcharge. (That wasn't a recurring expense and should not have been included in overhead.) I've connected the dots at the 16 percent markup and 16.5 percent markup points to create the short dashed line that projects upward from left to right.

The two overhead figures we've used, $382,000 and $420,000 respectively, represent 7.8 and 7 percent of gross receipts. Let's assume an average overhead percentage of 7.5 percent. The *Calculated overhead* line in Figure 10-6 is 7.5 percent of volume at each of the markup percentages shown.

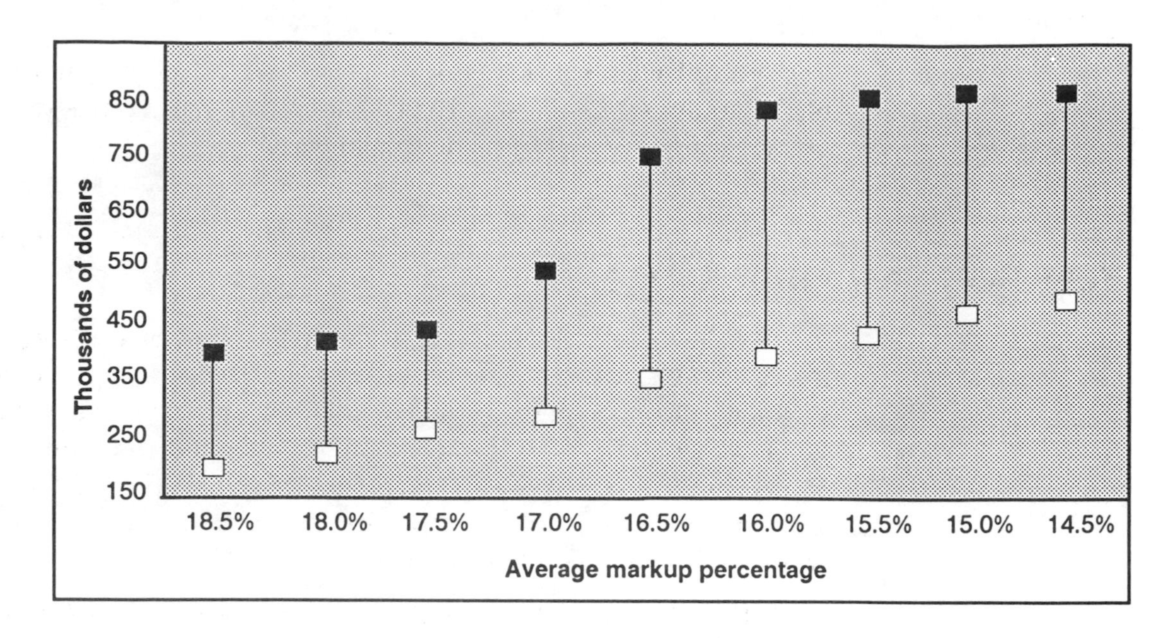

Figure 10-7 Net profit

You can see that the dotted line connecting the ends of the *Calculated overhead* line follows the line we laid out for overhead at 16 percent and 16.5 percent. That's confirmation that calculated overhead is charted correctly.

The space between the *Gross profit* line and the *Calculated overhead* line show the net profit. Let's simplify this chart a little to concentrate on net profit.

Net Profit

Figure 10-7 is the *Net profit* graph. The space between the gross profit marks on the top and the overhead marks at the bottom shows net profit. The bigger the space between those lines, the greater the profit. At some point beyond the right side of the profit vs. overhead graph, the lines connecting the marks will cross. That's the company's break-even point. Anywhere beyond that, the company will lose money because overhead will be more than gross profit.

There's a point in the middle of the graph where you make the most money. You want to set prices where the spread between the dots on Figure 10-7 is the greatest. That's not necessarily at the point of greatest volume, or even greatest gross profit. It's the point where there's the greatest difference between gross profit and overhead.

AEI Builders' best profit area is somewhere in the 15-1/2 to 16 percent range. If AEI Builders moves to a slightly lower markup (16 percent instead of 16.5 percent), their net profit before taxes will increase.

Therefore, 16 percent is what we'll pick for AEI Builders' normal business plan markup. AEI will use a 16 percent normal markup unless there's a good reason to deviate from those numbers.

Your Own Best Profit Range

Use the same method to identify your own best markup. All you need is a history of bids you've won and lost, and company expense records. If you don't have complete records of winning and losing bid totals, use what you have and try to collect more complete information on future jobs. Most owners are anxious to have contractors submit bids. I've found that owners will usually commit to providing information on the winning bid if you request it before you submit your bid.

What if you don't have a bid history? Easy. You can estimate your own best profit margin without a detailed bid history. Just analyze your financial statements for the past two to three years. At what markup did you earn the biggest net profit? Of course, your profit margin can vary as the construction economy varies. Your profit this year may be better than last year because more jobs are available now. It's hard to isolate the effect of changes you make in pricing strategy.

No matter what your strategy, you're not going to cut markup by one-half, or double it. At the most, you'll fine-tune your markup by a percentage point or two. In our example, AEI Builders is already in the high profit range for today's marketplace. We didn't even consider cutting the markup in half or doubling it. Any change of more than 2 percent puts them in danger of losing money. We're betting that a very minor reduction in markup will be favorable to the bottom line.

Smart Bidding

Of course, there's more to winning jobs than just picking the right markup percentage. What we've learned from this chapter is the starting point. Now let's look at some reasons why markup should vary from the standard you've identified. That's the subject of Chapter 11.

Chapter 11
Smart Bidding

As I explained in the last chapter, there's some (normal) markup that's right for your company. But that's only the average markup. Sometimes you want more than average markup and sometimes you'll have to settle for less. Here are some reasons why you'll vary markup from job to job:

- Adjustment for risk
- Asset utilization
- Market-based pricing

Any of these factors might justify changing your normal markup for a particular job. I'm going to cover the first two in this chapter. In Chapter 12 I'll zero in on market-based pricing — how to scope out your competition.

Adjustment for Risk

In Chapter 10 we calculated the normal markup for AEI Builders at 16 percent. Let's look at some reasons why this company (and yours) might use a markup different from what's prescribed in the company business plan.

Risk analysis

What's the probability that any particular job will cost more than you estimate it will? And how much more? The way you answer those questions is what we call *risk analysis*. If you're bidding what you consider to be a high-risk project, I recommend you use a higher-than-normal markup to protect yourself.

Here are some things that add risk to your jobs:

Weather

How do you build weather conditions into your estimates? Once you know how long a job should take under ideal conditions, you can predict with some accuracy how much additional time to allow for bad weather.

Suppose you're bidding a job in July that's due to begin in November. Your schedule calls for 10 months, or 220 work days, of construction time. In a typical winter month at this job site you can expect to lose an average of six days to bad weather. So, over the first four months you can expect to lose around 24 productive work days. Build that extra time into your construction schedule.

Hazardous Work

This should already be built into your labor rates. For instance, your labor rate for roofing should include the Workers' Compensation premium for that trade. That's part of your normal contractor's insurance and tax burden.

But consider whether anything on site may reduce productivity below normal levels. In earlier chapters I emphasized the importance of the job site visit. Do a little evaluation early in the estimating process. Then ask yourself some important questions before you decide on the profit margin. What's the neighborhood like? Will you have to provide extra security? Will traffic congestion around the site slow you down?

Cost Overruns

This is a risk you face on every contract. What are the chances this job will cost more than the estimator says it will? Answer this question by examining cost records and estimates of completed jobs.

Use project summary sheets

I recommend you fill out a project summary sheet like Figure 11-1 for every job. Then accumulate information from summary sheets to a completed project worksheet like Figure 11-2. Notice that Figure 11-2 lists costs from all AEI jobs for an entire year. They're the same jobs as the bold ones in Figure 10-1, the bid history (in the previous chapter). Use the blank worksheet, Figure 11-3, to make your own cost comparisons.

Figure 11-3 will be more useful if you make comparisons by categories. For example, if you do only tenant improvement work, divide your projects between offices and warehouses, or between residential and commercial projects.

Project Summary

Job number ____________ Estimator ____________

Project ____________ Project Manager ____________

Owner ____________ Location ____________

Location ____________

	Labor	Material	Equipment	Subcontract	Total
Actual cost					
Estimate					

Overrun					

Bid amount ____________
Estimated cost ____________
Projected profit ____________
Cost overrun ____________
Bid profit ____________
Change orders ____________
Cost of changes ____________
Total profit ____________

Figure 11-1 Project summary form

This form has columns for labor, material, equipment, and subcontract costs. Break your costs down by trade to help you pinpoint exactly where estimated costs don't match actual costs. Some trades present more estimating risk for your company and should carry a higher markup.

In Figure 11-2, the figures in bold type are the actual costs for each part of the job. The number to the right of the bold figure in each column is the estimated cost for each construction element.

Begin with job #401 on the top line. You can see the estimated time to complete this job was 70 working days, and that's how long it took. AEI Builders estimated labor at $101,000, but the actual cost was $98,000. Material costs were estimated at $60,000 and actually cost $59,000. They paid $101,000 to subcontractors and had estimated this cost at $94,000. Equipment costs wound up as expected, but the total overrun for the job was $3,000.

When you divide the overrun by the total estimated costs, you see that the overrun was just a bit over 1 percent.

✏ $3000 ÷ $269,000 = 0.011, or 1.1%

Offices

Job #	Work days		Labor costs		Material costs		Equipment costs		Sub costs		Total costs		Change orders
401	**70**	70	**98,000**	101,000	**59,000**	60,000	**14,000**	14,000	**101,000**	94,000	**272,000**	269,000	0
406	**70**	70	**95,400**	95,000	**65,000**	65,000	**14,000**	14,000	**88,000**	87,000	**262,400**	261,000	0
408	**78**	80	**106,800**	105,000	**83,000**	85,000	**21,000**	21,000	**144,000**	144,000	**354,800**	355,000	0
409	**182**	160	**273,300**	260,000	**250,000**	240,000	**97,000**	90,000	**427,000**	425,000	**1,047,700**	1,015,000	0
Totals			**573,500**	561,000	**457,400**	450,000	**146,000**	139,000	**760,000**	750,000	**1,936,900**	1,900,000	0
Over/Under			12,500		7,400		7,000		10,000		36,900		

Apartments

Job#	Work days		Labor costs		Material costs		Equipment costs		Sub costs		Total costs		Change orders
403	**168**	160	**169,300**	170,000	**224,000**	225,000	**124,000**	120,000	**543,000**	540,000	**1,060,300**	1,055,000	5,000 Sub
404	**142**	150	**112,000**	110,000	**132,000**	150,000	**92,000**	95,000	**418,000**	418,000	**774,000**	773,000	
Totals			**281,300**	280,000	**356,000**	375,000	**216,000**	215,000	**961,000**	958,000	**1,834,300**	1,828,000	5,000
Over/Under			1,300			-19,000	1,000		3,000		6,300		

Houses

Job#	Work days		Labor costs		Material costs		Equipment costs		Sub costs		Total costs		Change orders
402	**58**	55	**18,700**	18,000	**41,000**	40,000	**0**	0	**39,000**	39,000	**98,700**	97,000	1,500 Mat'l
405	**60**	55	**21,500**	19,000	**49,400**	48,000	**0**	0	**44,000**	43,000	**114,900**	110,000	500 Mat'l 500 Labor
407	**56**	50	**19,000**	17,000	**34,000**	33,000	**0**	0	**41,000**	40,000	**94,000**	90,000	
410	**66**	65	**25,500**	23,000	**60,000**	60,000	**0**	0	**62,000**	62,000	**147,500**	145,000	3,200 Mat'l
Totals			**84,700**	77,000	**184,400**	181,000	**0**	0	**186,000**	184,000	**455,100**	442,000	5,700
Over/Under			7,700		3,400				2,000		13,100		
Grand total			**939,500**	918,000	**997,800**	1,006,000	**362,000**	354,000	**1,907,000**	1,892,000	**4,226,300**	4,170,000	10,700
Over/Under			21,500			-8,200	8,000		15,000		56,300		

Figure 11-2 AEI completed projects worksheet

Job #	Work days A	Work days E	Labor costs Actual	Labor costs Est	Material costs Actual	Material costs Est	Equipment costs Actual	Equipment costs Est	Sub costs Actual	Sub costs Est	Total costs Actual	Total costs Est	Change orders	For:
Grand Total														
Over/Under														

Figure 11-3 Project worksheet

Calculate Cost Overruns

Look at the Completed Projects Survey for AEI Builders in Figure 11-2. Use that information to figure the cost overrun percentage for each of the following categories:

Percentage overrun by project type	
Office buildings	______ %
Apartments	______ %
Houses	______ %
Percentage overrun by cost element	
Labor	______ %
Material	______ %
Equipment	______ %
Subcontract	______ %
Percentage overrun by project size	
Small (under $200,000)	______ %
Medium ($200,000 to $1,000,000)	______ %
Large (over $1,000,000)	______ %
Percentage overrun for all jobs	______ %

Figure 11-4 Cost overrun exercise

To see exactly how this works, complete Figure 11-4. In the first section, find the cost overrun for each of the project types. Subtract the actual cost from the estimated cost. Then divide the difference by the estimated cost.

For offices, the calculation is:

✏ $1,936,900 - $1,900,000 = $36,900

✏ $36,900 ÷ $1,900,000 = 0.019, or 1.9%

Fill out the rest of this exercise yourself. Answers are in Figure 11-5.

Notice that you use the total dollar volume in each category, not the average overrun for all jobs in the category. Here's why. Suppose you did two jobs last year. On one of them your estimated cost was $100,000. On the other it was $1,000,000. Also, suppose your overrun on each job was $10,000. Your overrun was 10 percent on the first and 1 percent on the other, for an average of 5.5 percent. But that's not an accurate indicator of your estimating accuracy:

✏ $1,100,000 x 0.055 = $60,500

Your overrun for the year was actually just 1.8 percent of gross:

✏ $20,000 ÷ $1,100,000 = 0.01818

And this proves to be true:

✏ $1,100,000 x 0.0182 = $20,020

Calculate Cost Overruns

Look at the Completed Projects Survey for AEI Builders in Figure 11-2. Use that information to figure the cost overrun percentage for each of the following categories:

Percentage overrun by project type	
Office buildings	1.9%
Apartments	0.3%
Houses	3.0%
Percentage overrun by cost element	
Labor	2.3%
Material	1.2%
Equipment	2.3%
Subcontract	0.8%
Percentage overrun by project size	
Small (under $200,000)	3.0%
Medium ($200,000 to 1,000,000)	0.3%
Large (over $1,000,000)	1.8%
Percentage overrun for all jobs	1.35%

Figure 11-5 Cost overrun solution

And notice the bottom line. The estimated cost for all types of work in Figure 11-2 was $4.170 million. The total overrun for the year was $56,300, or 1.35 percent.

Knowing the usual "miss" on estimates should help you make more accurate estimates in the future. Notice that comparing actual and estimated costs is possible only if you keep accurate records of completed jobs. That's important. Without actual costs, there's no way to check estimating accuracy. You know exactly how much you estimated each job would cost. You *should* know exactly how much each job actually *did* cost. If you need more suggestions for keeping good cost records, I can recommend *Cost Records for Construction Estimating* by W.P. Jackson (published by Craftsman Book Company). An order form is at the back of this book.

A "miss" of 1 to 2 percent may be acceptable

What's an acceptable overrun? - Are your estimates as accurate as the estimates in Figure 11-2? If your average miss is under 2 percent, is that close enough? If your average miss is more than 2 percent, is that about right for your kind of work? Is it too high? Or, do you wish you could do better?

Of course, every estimator wants the smallest "miss" possible. An average miss of zero would be perfect. But it isn't necessary or even desirable. For most contractors, a miss of 1 or 2 percent is "plenty good enough." It takes too much time to estimate more accurately than that. It's foolish to spend 10 estimating hours finding the last $50 of job cost.

Of course, your average "miss" will vary with the type of work you're estimating. Custom home builders who subcontract nearly all their work should be accurate to within 1 or 2 percent, especially on tracts where they build from the same plans over and over again. Remodeling contractors won't do nearly as well. Accuracy to within 8 or 10 percent may be close enough. Expect overruns and allow for them in your estimates.

There are many more estimating unknowns and unexpected costs in remodeling, repair and renovation work. There's no way to be sure what's required until you open up the wall and start cutting out studs. There's no way to know ahead of time what it's going to take to please the owner. Until the owner is satisfied and has paid the bill, your work isn't done.

The average miss for AEI Builders (1.35 percent) is good enough. They do mostly new construction and use their own crews for about half their work. Your estimates on subcontracted work will usually be more accurate than estimates on work you do with your own crews. If not, you're either using the wrong subs, or not watching the scope of work included in their bids.

Now, look back at Figure 11-5. The first breakdown by project type may not tell you anything significant. There are too many variables to consider when a company builds vastly different project types. The product mix in your company will probably be much more specialized. That makes it easier to analyze your overruns.

Risk by Cost Element

Now look at the middle category, the breakdown by type of cost. These figures are much more revealing. You can expect the greatest overruns to be in labor cost (and equipment cost, if you have company-owned equipment). Material cost overruns should be much less. Overruns on subcontracted work should be the least of all.

Subcontracts offer the least risk

Generally, subcontracts offer the least risk. If a subcontractor misses something in an estimate, it's their problem. They have to perform at the contract price.

But if you omit something on work done by your crews, it's *your* loss. And there's another risk when you estimate work to be done by your own crews. Suppose you misjudge the complexity of a job. Instead of framing 12 linear feet of form per hour, soil conditions reduce productivity to 6 linear feet per hour. That's a major problem on a job that includes hundreds of feet of forms.

Any job can be harder than you expect. And job difficulty isn't the only thing that affects completion time. Crew productivity depends partly on the task itself, partly on the crew that's doing the work, and partly on the working conditions. That's a lot to consider. It takes experience to judge which jobs will go faster or slower than normal.

Labor costs will always be higher than normal on more complex jobs with marginal working conditions (such as bad weather) and low crew productivity (due, for example, to poor supervision). We all know that. But no estimator can be expected to anticipate all the problems that can influence installation costs.

Material costs are different. They don't change as much. And it's unlikely you'll use twice as much material as expected (unless your crew has to tear something out and start over again).

Risk by Job Size

I think you'll agree that labor cost is the high-risk estimating category in the second section of Figure 11-5. In the breakdown by project size, your own numbers should also be about like our example. Your dollar ranges for small, medium, and large projects may be different from ours. But expect your greatest cost overruns to be on smaller jobs. It's easy to figure that a little job will take five days and then spend six on it. That's a 20 percent labor cost overrun.

Medium-sized estimates will be the most accurate

Estimates for the medium-sized jobs, the ones you do most, will probably be most accurate. Those should be the jobs you bid and manage best. They shouldn't give you many surprises. Estimates for bigger jobs probably won't be as accurate. The jobs you bid once a year are a tougher test of your estimating skill. Most estimators bid an unusually large project very carefully. But cost overruns are still more likely on these jobs than the modest-sized jobs you handle routinely.

What If Your Pattern Is Out of Line?

What if cost overruns are worse for subcontract estimates than for your own labor estimates? What's wrong? A couple of things could cause that. You may not be paying enough attention to the scope of work, or your subcontractors' understanding of it. Or maybe you're too lenient about approving extras in the field and not charging them back to customers. In any case, subcontract overruns should be much less than labor overruns.

This analysis may reveal more about your crews than about your estimates. In any case, use it as a tool to identify problems that are sucking dollars out of your profit margin.

Here's a caution: Beware of focusing on overruns on any single job. Managers tend to get very excited about a single project that goes way over estimate. But your primary focus should be on the overall miss, not the worst case. The goal is to be on target most of the time. Don't get too concerned about a single bad job.

Analyze the Cost Overrun Breakdown

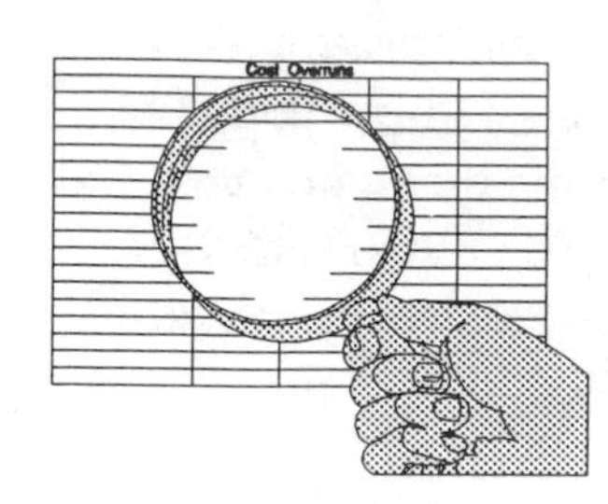

Is any type of work shown in Figure 11-5 a particularly high-risk category for AEI Builders? No, not at first glance. Their average overrun is 1.35 percent, ranging from 0.3 percent (which is next to nothing), to 3 percent. None of AEI Builders' work categories show a high financial risk. Neither is the company losing money on any work category.

Earlier I mentioned that AEI Builders' overhead is 8 percent of cost. Their business plan calls for a 16 percent markup, which includes the 8 percent for overhead. What if they had one category where their overruns averaged 9 percent? At that rate, they would lose money every time they did that kind of work. The more work they did, the more money they'd lose.

That's not a problem for AEI Builders. But if you're losing money on a particular class of work, either stop bidding on that class or make some corrections — such as raising your markup or altering your bidding procedure.

Look for a Pattern of Errors

I've already said none of AEI Builders' work shows an obvious pattern of bad estimates. But look again. See if you can find a subtle problem lurking here. It's not in the "apartments" category. The miss there was only about 0.3 percent on $1.8 million in work completed. And it's probably not in offices, either. No single job missed by more than 2.2 percent.

Now look at the labor column for houses, back in Figure 11-2. There's a $7,700 overrun on $77,000 in estimated costs. That's a full 10 percent miss. And actual labor costs on all four jobs were more than estimated labor costs, from 3 to 13 percent.

Find unusual overruns

If this happens in your company, start asking some questions. I suspect someone in AEI Builders is too optimistic about labor costs on work done with AEI crews. And there's probably a reason for this. Maybe AEI Builders wants to use company crews rather than subcontract this work to a more efficient specialist. If estimates were more accurate, subcontract bids would probably come in below the estimated costs. And note this difference. Supervision expense is usually higher on work done with your own crews. Why waste time overseeing a framing crew, for instance? Instead, sub out the framing and use the supervisors to manage another job or two.

Make this comparison with your own figures. You may discover a type of work where you consistently overestimate costs — actual costs come in under bid prices. If that's the case, keep doing what you're doing. Don't sub out carpentry, for instance, if you always make good money using your own crew.

Use your own crew for the right reasons - Many contractors estimate the labor costs for company crews at the same price a subcontractor would have charged. That's fine. But be sure to keep accurate records of your labor costs. Was the actual labor cost more or less than the estimated cost? Or did it in fact cost the same as a subcontractor would have charged? If it did, is the additional control you get from using company crews worth the extra supervision time and trouble?

Know when to hire a subcontractor

When using your own crews, do it for the right reasons. If your crews are very efficient and make money for you, that's a good reason. If you're doing a job out of town where you don't know the subs, that's also a good reason. But, don't be like an estimator friend of mine. He told me, "Dad started this business 35 years ago as a carpentry contractor. We've always done our own carpentry." That's not a good reason if you lose money on most jobs.

If you're a builder who does some of your own specialty work, you're really in several different businesses. That's ok if you're making money at all of them. If you need to handle a particular type of specialty work because no subcontractor can do it the way you want, that's ok too. Just be sure you charge enough.

Why are you handling subcontract specialties with your own crews? Maybe you could make more money handling a higher volume of work as a general contractor. Consider giving up the small profit that's available from each of the specialty trades, and look for a bigger payoff elsewhere.

For AEI Builders, the 10 percent underestimate on labor for houses is too much. It's clearly high-risk estimating for AEI Builders. Every dollar they spend on labor for house construction costs AEI a dollar and two cents. Remember, their markup averaged 16 percent and overhead was 8 percent. So a 10 percent overrun left a 2 percent loss.

How Does Management React?

Two possible solutions

AEI Builders has two choices:

1) Bid labor for single-family homes at cost plus 16 percent, plus an additional 10 percent to cover the overruns.

2) Find the reason for the overruns and correct it.

The first choice is the easiest. But the risk is that bids will be too high. Volume will fall off. That leaves the second choice as the only practical alternative. Do estimates reflect too much optimism about productivity? Or is productivity just too low? Or is the supervision poor? First, you have to find out what the problem is. Then you have to correct it. In the meantime, adjust estimates so they're more realistic.

You May Need Other Breakdowns

Your analysis may be different

So far we've analyzed actual vs. estimated costs three ways:

- By type of project
- By element of construction cost
- By project size

Of course, there are other ways to group costs. Do you handle many jobs for the same customer? If so, compare results for that customer with results for other customers. Do you build a significant amount of work out of town? How do those jobs compare to the jobs you do locally? You'll probably find the local jobs are more profitable or require a lower markup. You're probably more familiar with local building department requirements and know what to expect from the inspectors. And local jobs have lower mobilization costs.

If you have several supervisors, is one supervisor bringing in profits while another produces losses? Don't let a single bad job obscure the picture. What we're looking for here is a pattern of bad results. Do a couple of your supervisors always bring their jobs in under cost, or within a percent or so? Is there one supervisor who runs over 6 or 7 percent on most jobs? Maybe it's not just random chance. If there's a pattern you can identify, take steps to correct it.

Group your jobs by category in any way that yields an accurate picture of company performance. The point is this: It's not enough to keep good cost records. That's just the beginning. You have to use that information to find problems and fix them before they destroy your company.

Asset Utilization Adjustments

Some construction projects are right for your company and some aren't. No construction contractor can bid on all the jobs available. The projects you should bid on depend on the assets you have available. Let's examine the assets your company has (and a few it may not have).

Construction companies generally have four kinds of assets. These aren't necessarily assets that appear on your financial statement. These assets limit how much work your company can reasonably expect to do, the type of work your company should be doing, and how you should be scheduling that work. Production assets fall into these categories:

Four types of production assets

- Equipment
- Supervisors
- Labor force
- Cash

All jobs require some of each asset category. Most companies don't come close to fully using all assets all the time. They can't. As a practical matter, when one of the four asset categories is fully committed, your company is probably tapped out. You can't take on any more work, even if assets are available in other categories.

For example, if your working capital (cash) is fully committed to work in progress, it's unwise to take on more work, even if more equipment, supervisors and manpower are available.

Of course, you can usually buy or rent more equipment, hire more supervisors, recruit more workers or borrow more cash. But that takes time (anywhere from a few days to a few months) and comes at a cost. For the immediate future, there are practical limits to the assets you have available.

Here's why the concept of asset utilization is important. I recommend that you adjust prices based on asset utilization. The markup you add should depend on the following:

Factors affecting markup

1) How much work is available?
2) How big is the job?
3) When can you expect to be paid?
4) How busy are your crews and supervisors?
5) Do you have equipment sitting idle?

Let's examine each of these points in detail to see how each should affect markup.

What's the Market Like?

This point should be instinctive. If the market is depressed, trim your markup to stay competitive. Be sure estimates cover overhead, but be willing to accept less profit than usual. Your competition will be doing the same thing. Every contractor does that in a weak market. When every contractor is working six days a week, you can use a fatter markup and still get plenty of work.

Job Size

AEI Builders' jobs range in price from around $100,000 to just over $1,000,000. Should markup be the same on all these jobs, regardless of size? Probably not. Most contractors don't use the same markup on a large job as they do on a small one.

Your own job size range may be very different from AEI Builders. But you still have low end (small) jobs and high end (large) jobs. Large jobs are always worth more to your company than small jobs. That's because larger jobs use company resources more efficiently. You need a lower percentage of job revenue to cover overhead. So your markup on a $1,000,000 job might be less than 16 percent. For a $125,000 job, markup would have to be higher. The question is, how much lower or how much higher? We'll cover that later in the section on project overhead adjustments.

Look for Favorable Cash Flow

Suppose you're bidding two jobs and they're fairly equal in size and scope. Both will take about 90 days to complete. But on one of them you'll send bills weekly and expect to receive progress payments within five days after you submit each invoice. On the other job, payments will be a problem. There are only two progress payments. The last one-third of the contract price isn't due until completion.

Unless your bank balance is fatter than mine, there should be no question about which job gets the lower markup. Your workers and suppliers won't be willing to wait for their money. You'll have to charge more if you're willing to carry this job yourself.

By the way, this strategy works with your subcontractors also. Build a reputation for paying subs regularly and promptly. You'll probably get both better prices and more cooperation when you need a favor.

Identify Under-utilization

To do the next exercise, you'll have to refer back to Figures 9-2, 9-3 and 9-4 in Chapter 9. These are the company profile and financial statements for AEI Builders. You'll also need to refer back to Figure 11-2, the completed project worksheet for AEI Builders, in this chapter.

Examine the company profile, balance sheet and income statement and find at least three ways that AEI could make better use of construction assets. In each case, calculate the increase in sales and profit that optimum use of each asset would produce. Pay special attention to the company profile. It has important information you need to complete this exercise.

Remember, these aren't necessarily the same assets you'll see listed on a balance sheet. Still, they're real assets. Without them AEI couldn't complete work and earn a profit. In many ways, these assets are as important as cash in the bank. They're more important than the office you work out of.

On Figure 11-6 list each asset you think AEI Builders didn't use to its full potential. Beside each item on that list, write down how much more work AEI Builders could have sold if they had used each asset completely. Then make an estimate of how much extra profit AEI would have made by

Maximum Asset Utilization

Asset	Revenue	Profit

Figure 11-6 Asset utilization exercise

full use of their assets. You don't have enough information to assign a dollar value in every case. But at least list each asset the company didn't use effectively.

Notice that we're working with ideal situations here. In practice, maximum utilization of all assets is probably impossible. But for this exercise, assume perfect conditions without limitations. Calculate the additional profit that was possible from the information you have available.

When you're finished, look at Figure 11-7. That's my solution. Next, I'll explain how I developed these answers.

Maximum Asset Utilization Asset	Revenue	Profit
Equipment	238,000	Slightly
	to	less than
	2,975,000	238,000
Managers	1,126,000	146,380
Superintendents	?	?
Cash	340,000	47,000
Office space, yard storage	?	?
Crews, supervisors	?	?

Figure 11-7 Asset utilization answers

Under-use of Equipment

This is the most obvious under-utilized asset. Bear in mind that we're not just looking to break even on company equipment. We're trying to find a way to increase the use of this asset category and make money on it.

If AEI Builders kept all their equipment working all year, full time, how much could they increase sales and profit? The company profile says (in the third paragraph) that they could bill $50,000 per month to jobs, if all the equipment was in use all the time. That's a total of $600,000 per year. Note on the income statement that they actually billed $362,000. The difference is $238,000. If they had charged that much more to jobs, how much more profit would they have made?

Most equipment costs are fixed costs. Those costs go on whether the equipment is operating all day or sitting idle in your yard. Fixed costs include interest payments on loans, business insurance, depreciation, storage space, and some maintenance expense. Variable costs include fuel and maintenance and are proportional to the use of equipment. The more you use your equipment, the more you pay for fuel and maintenance.

Here's a key point. Recognize that most of the $238,000 that could have been billed for equipment would have gone straight into profit. The only added cost would have been fuel and maintenance. We know revenue will increase by at least $238,000, but it could be a great deal more. In fact, the equipment available could have handled an extra $3 million in work if equipment expense is 8 percent of total costs. That's an example of under-utilization in spades.

Management Personnel

The company profile says each project manager/estimator can handle about $3 million in work each year. The company did $4,874,000 for the year in our example. That means those two managers could have handled another $1.126 million. We know that AEI Builders earned 13 percent gross profit from construction. That means they could have expected to make another $146,380 (less overhead) from the added $1.126 million in gross sales.

There's a flip side to the discussion of project managers. What do you do in an especially slow year? Suppose you're paying project managers $50,000 a year and they're busy only half the time? Of course, you want to keep them on the payroll if you can. It takes at least a year to train a good project manager/estimator. You don't want to lose them to a competitor. And certainly you'll provide a bonus in years when a manager/estimator produces up to or even beyond full capacity. You make a lot more money when a manager produces $3 million in a year than if the same manager produces $2 million in the same period.

Field Personnel

Here's another area where it's hard to identify the amount of unused capacity. Also, we don't know how much AEI Builders pays their crews or how many workers they have on the payroll. We do know they have a fairly large field force, and four superintendents. It seems likely to me that the four superintendents plus two project managers are more than adequate to handle ten jobs. My guess is that they could handle a lot more work without increasing payroll. You may agree, but we don't have enough information to know for sure. This is largely a matter of personal opinion and experience.

Description	Original	Adjusted	Difference
Gross sales	4,874,000	5,258,200	384,200
Construction costs	4,237,000	4,574,634	337,634
Gross profit	637,000	683,566	46,566
CD income	34,000		
Adjusted gross	671,000	683,566	12,566
Overhead	496,000	496,000	
Before tax profit	175,000	187,566	12,566

Figure 11-8 Investment vs. job financing comparison

Cash

Notice the line on AEI Builders' income statement that shows the interest on their CD's (certificates of deposit). How much money would you need in CD's to earn $34,000 in interest? When we compiled the figures for this sample company, CD rates were close to 10 percent. So it looks like AEI Builders was earning interest on $340,000 that could have been used to finance more jobs. Figure 11-8 shows that AEI Builders could have added nearly $47,000 to gross profits, and almost $13,000 to their pre-tax profits by using the $340,000 to increase total volume instead of investing it in CD's.

Some dollars are worth more than others - Here's something else to consider about cash. The last sales dollars each year are more important to your bottom line than the first sales dollars. You have to do a certain amount of business each year just to cover overhead. Once overhead expense is covered, profit per dollar of sales soars so long as markup remains the same.

Fixed overhead changes in steps

Notice that we base these calculations on gross profit from construction, not the net profit before taxes. Overhead costs are part of the net income calculation on your income statement. If you use the net profit ratio, you're saying that office expense and management salaries will rise proportionately to gross volume. That isn't usually true. Fixed overhead tends to change in steps rather than rise and fall smoothly with volume. Usually you can increase volume 10 or 20 percent with little or no change in overhead. Office rent, utilities, and management salaries will probably stay about the same whether you do $5 million or $6 million in a year. Even if you double volume in a year, overhead probably won't double.

Here's another way to look at the effect of volume changes on net income. Figure 11-9 is what the income statement for AEI Builders would have looked like if they had done only $4 million for the year. You can see there would have been virtually no profit, and perhaps even a loss.

AEI Builders Income Statement		
Gross sales		$4,000,000
Construction costs		
Labor	772,000	
Material	840,000	
Equipment	297,000	
Subcontracts	1,571,000	
Total costs		-3,480,000
Gross profit from construction		520,000
Other income: Interest from CD's		0
Adjusted gross income		$520,000
Overhead		
Management salary	80,000	
Estimator/Project Manager salaries	110,000	
Clerical salaries	96,000	
Office expense	48,000	
Legal, audit, miscellaneous	48,000	
Equipment not costed to jobs as expected	114,000	
Total overhead costs		$-496,000
Profit before taxes		$24,000

Figure 11-9 Income statement at $4 million gross

Overhead costs remain the same. Gross profit barely covered overhead. (We've assumed the company didn't receive any interest on the CD investments.) Note also that equipment cost would have been an even greater burden at a smaller business volume.

Once you've covered overhead, the markup on everything else you bill is almost entirely profit.

Company Facilities

AEI Builders owns their own building, but the company profile tells us they have more space than they need. Surplus space isn't a major under-used asset. But extra office, warehouse and yard space could bring in additional revenue. If you have extra space, fence it off and offer it to one of your subs at a reasonable rent. The extra income is almost pure profit.

Can you see other indications that AEI Builders isn't using company assets to best advantage? Notice that they have 30 to 40 workers in the field and four superintendents. They did 10 jobs during the year, requiring a total of 950 work days. And 45 percent of their costs were for subcontractors. From this information I would guess that the supers are probably handling only one job at a time. I also suspect that field crews aren't working at peak efficiency.

Set productivity goals

Our point here is this. You need to set productivity goals for crews and superintendents. Then monitor productivity to be sure you get what you pay for. The best supervisors do this very well, without making workers feel pressured. Several good books of manhour standards are available. One of the better values is the *National Construction Estimator* published by Craftsman Book Company. You can order a copy using the order form at the back of this book.

Project Adjustments

Now we know the amount that better utilization of assets could add to profit, what changes should we make? Let's concentrate on equipment and manager/estimator utilization, since those seem to be the assets with the most potential savings. We'll adjust bid prices to make better use of those assets.

I recommend you make these adjustments in dollars, not percentages. In our earlier discussion, we talked about markup in percentages. We decided on a business plan markup of 16 percent. AEI Builders needed another 10 percent adjustment for higher-than-expected labor costs on single-family houses. Now we'll adjust to maximize asset utilization in dollars — such as "Add $5,500 to this bid," or "Deduct $20,000 from that job."

Make adjustments in dollars

So, how do we decide on the amount of the adjustment? Look at the maximum amount of work you could do in each category. AEI Builders has project managers who can handle $6 million in a year. They have equipment they can bill out for $600,000 in a year. Of course, AEI's estimators can't plan each job as though all company equipment is available for that job at all times. They have to know how much is already committed to other jobs and what will be available for the job they're bidding.

Let's summarize what we already know about use of company manager/estimators and use of company equipment.

1) Our goal for this year is $6 million in business, based on manager availability. We're going to keep them busy full time this year.

2) Last year, AEI Builders billed $362,000 in equipment charges, and charged back $114,000 to overhead. This year, our budget says we'll eliminate the back charge. So it follows that our budget calls for roughly $480,000 in equipment billing this year to break even.

3) If we keep our equipment busy all the time, we can bill jobs for the full $600,000.

From points two and three above we can calculate that we need 80 percent utilization of company-owned equipment ($480,000 ÷ $600,000 is 0.8). Most contractors can't expect to keep their equipment busy 100 percent of the time. Instead, they make some reasonable assumption about usage. For our purposes, we're going to assume 80 percent utilization for equipment and 100 percent utilization for project manager/estimators.

Project Adjustment in a Nutshell

- If a project will put unused assets to work, it's good for the company. You can afford to reduce bid prices because of the financial benefit to the company.
- If this project will leave company assets unused, you'll have to increase markup beyond what you'd usually charge.

Cut your price when... Suppose you're chief estimator at AEI Builders. You've prepared a bid with a price of $1 million. You've bid this job with the normal business plan markup of 16 percent. The project will take two months to complete and this is the only job you'll be working on during that two-month period. So far this fits your plan exactly. Your budget calls for $6 million for the year. This job accounts for one-sixth of that total.

But there's a lot of equipment cost in this project, $95,000 to be exact. In other words, you plan to charge the job $95,000 for company-owned equipment for the two months the equipment portion of this job will take.

Equipment earns more than its share

Your goal for the year is to bill $480,000 in equipment, an average of $40,000 per month, or $80,000 for two months. In two months this job will return $95,000, or $15,000 more than the goal. You can afford to reduce your bid price on this job by $15,000 and still come out on target when the job is done.

Here's how that works. Suppose you get the job. The internal billing cost for equipment at the end of the job is $95,000. Your budget calls for billings of $80,000 during that period. So you have a windfall of $15,000 profit from the use of equipment. You can afford to reduce profit margin on the job by $15,000 because you've gained better utilization of equipment by $15,000. Reducing the bid price by $15,000 increases your chance of getting the work.

But raise your price if... Sometimes it works the other way. You have to increase the profit margin because a job would result in under-use of assets.

Suppose I'm your customer. You've handed me a bid on a custom home addition. The bid is $150,000 and assumes your normal markup. I open the bid and explain that I'm satisfied. But, I say, "I'll give you this job on one condition. I want you to assign your best project manager to this job exclusively for the entire construction period, one full month."

Whoa! This means you have to tie up a major asset — one of your two project managers — for a full month. A key construction asset, one of your project managers, will handle less than the normal workload for a full month. Remember, your project managers are expected to handle $3 million a year ($250,000 a month). A project manager doing only $150,000 in a month is under-utilized by $100,000. If you're willing to do that, you have to charge enough to make up for the overall reduction in gross income. How much more should you charge me?

Charge for special service

You'll have to bypass $100,000 in work for the one month project duration. If your business plan (normal) markup is 16 percent, you'll need to add $16,000 to the price of this job. That's what it costs you to meet the owner's demands.

We're assuming here that business is normal. If you don't get this job, there's enough work out there that you'll still meet your plan for the year. So your price for this job should be $166,000, not the $150,000 in your original bid. You're willing to do this job my way at that price. When it's completed, your profit will be the same as if your volume had been $250,000 for the month, at normal markup.

Try This for Yourself

Try an asset utilization exercise on your own. Suppose you're the chief estimator for AEI Builders. You're bidding a project with a total estimated cost of $1 million. The labor, materials and subcontract costs are $900,000. The cost for company-owned equipment included in the bid is $100,000. You estimate the job will take four months to complete. Your normal markup is 16 percent, so you expect to bid this job at $1,160,000.

Assume that this job, along with your other work in progress, will provide exactly the workload you need for the four-month period. Don't worry about the size of the job or your staff availability. Instead, focus on asset utilization. Should you increase markup because the job promotes under-utilization of equipment? Or can you cut markup because the job results in better utilization of equipment? Enter your answer on the blank line below before you read the next paragraph.

My answer is: ______________________________

If you suggested increasing the bid, you've concluded that this job will result in under-utilization of company assets. Let's see if that's really true.

First, compare this job to what your business plan says should happen during the time this job is running. If it uses a greater proportion of equipment, you can afford to cut the markup. If it uses less, leaving equipment sitting idle in the yard, you have to raise the markup. If it uses just about the right amount of equipment, you don't have to make any adjustment.

If you get this job at $1,160,000, what percentage is that of your total projected workload for the year?

✏ $1,160,000 ÷ $6,000,000 = 0.1933

This job represents 19.3 percent of your total workload. If it uses equipment at the rate you want it to, it should be 19.3 percent of your total equipment billing for the year.

✏ $480,000 x 0.193 = $92,640

Remember, estimated equipment expense was $100,000. Your conclusion: This job will absorb some of that excess equipment capacity, $7,360 to be exact ($100,000 minus $92,640 is $7,360). You can afford to lower the bid price by $7,360 without compromising your profit goal. Markup can be a little less because equipment billings will be a little more.

Note carefully that this example applies to your company even if you don't own expensive construction equipment. Use this same procedure to improve utilization of any type of asset: managers, cash in the bank, work crews, even estimators.

What Is the Net Effect?

Here's what's likely to happen when you adjust markup to improve asset utilization. You'll be more competitive on jobs that use company assets more efficiently. And you should get more of that type of work. You'll submit higher bids for jobs that don't use company assets efficiently. So you won't get as many of those jobs. The end result will be a workload better suited to your company.

This is exactly how managers in major engineering and construction companies tailor their bids to fit company requirements. If it works for the big boys, you should follow the same approach. Use the procedure I recommend. Lower your markup for big jobs and increase markup on smaller jobs. Bid a little lower when volume is down and a little higher when you're loaded. Favor jobs that make better use of company equipment and other assets. That should spell better earnings and more profit at the end of the year.

In a smaller construction company (under $1 million a year), I suggest that you use this concept of asset utilization on only one or two key assets. Don't bother working up formulas for all company assets. Follow your instincts on the less important assets. Concentrate on the key asset, which is probably project management capacity.

Moving On

The next (and last) topic will be competitor analysis — judging your competition. What are the other bidders going to charge for the job we're estimating today? Read the next chapter and you'll be able to make some very informed guesses based on what competitors have charged before and how busy they are right now.

Chapter 12
Pricing Strategies

No construction estimate is complete until you've considered the competition. Why worry about the other guys? Because your business depends on it. Ignore what other bidders are likely to do and you'll probably lose either some jobs or a healthy share of your profit.

There's no reason to lose a good job if you can cut a few dollars and get the work. And it's just as foolish to submit an estimate that's way below what the next lowest bidder is asking.

We call this *market-based* pricing. Every estimator should do it. Some call it competitor-analysis pricing or strategic pricing. All those names mean the same thing to me: beating the competition, but not by too much.

Three things market-based pricing does

Market-based pricing does three things:

1) It leaves less money on the table

2) It makes it harder for competition to predict your price

3) It turns more of your bids into contracts

Let's talk about the first expression, *"leaving money on the table."* Here's an example: Suppose you're about to submit a bid at $1,152,000. You're almost certain that none of the other bidders can bid less than $1.2 million. What will you do? That should be obvious. You'll bid the job at about $1,192,000, increasing your profit margin by $40,000. Failing to do so would be "leaving money on the table." You could have had the $40,000 in your pocket, but you left it on the table. Of course, that kind of opportunity doesn't come along every day. But when it does, you certainly want to know about it.

Item 2, making it harder for competition to predict your price, is much easier. Every contractor in a competitive market (I suspect that means every contractor) should be a little unpredictable. A construction company that's easy to predict will lose work to sharper competitors.

"Making educated guesses about predictable competitors can make or save you thousands . . . "

Suppose you're bidding head-to-head with another builder for a custom home project. You both know who's bidding the job. You know what the other builder's costs are. You also know that your most competitive builder always bids at cost plus 11 percent. If you really need the work right now, you can go in at cost plus 10.5 percent. You're likely to get the job. The other builder's predictability will cost him this work.

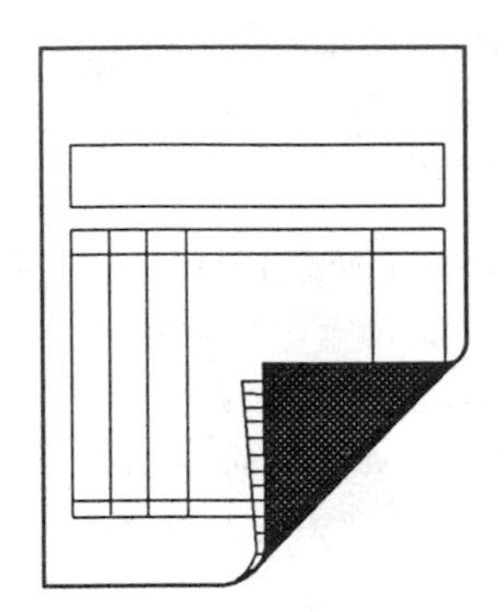

If you didn't know the other builder's habits, you couldn't take that job away from him. You'd have to do your best and take your chances. You might even bid at cost plus 9 percent. Making educated guesses about predictable competitors can make or save you thousands of dollars.

Market-based pricing is very serious business. True, it's less important if you get most work by negotiation. But even then, somebody out there is hoping for a shot at some of the (negotiated) action. Once they know exactly what you charge, it won't be too hard to work up an attractive bid.

Confuse your competition

Here's how to confuse the competition without hurting your business. When you'd rather pass on a job or are certain that the winner won't make much money, submit a bid higher than normal. It won't hurt your profit margin. You don't expect to get the job anyhow. And you'll throw the competition off track. I'll offer more examples later in this chapter.

The third reason for market-based pricing is to turn more bids into jobs. That's tougher. Everyone wants to raise their ratio of jobs won to jobs bid. All you need to do is predict in advance what it takes to get the work: a 1 or 2 percent discount? That's not easy to determine, even if you have lots of information on how the competition bids.

Most of the larger engineering and construction companies collect information on competitive bidders. Many use sophisticated analyzing systems to predict how a competitor will price a particular job. If you have a computer and know how to use a spreadsheet like Lotus 1-2-3 or Microsoft Excel, try this analysis on your computer. If you don't have a computer, no problem. Manual methods work almost as well.

Learn About Your Competition

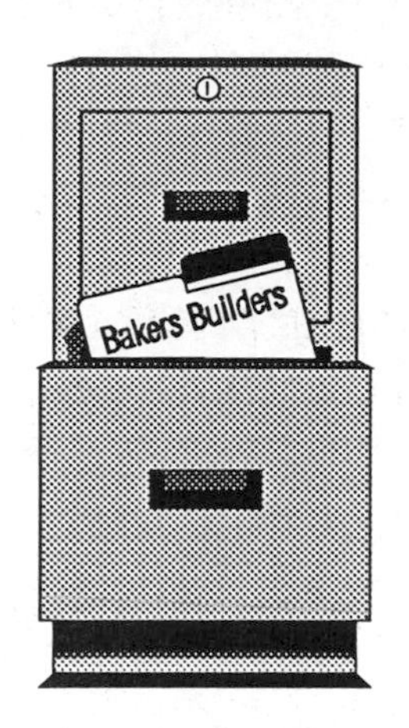

Set up a file for each of your regular competitors. Figure 12-1 is a blank form you can copy. Figures 12-2 through 12-7 are my examples. Use the information on competitor bid forms like these when you're bidding against one of your regular competitors. If you're bidding a job you really need and want, bid just a little lower than what you predict the competition will charge.

Now this assumes you *know* what the competition will charge. Is that possible? Yes! There's a reasonably accurate way to anticipate the bids of key competitors. First, add a line to the page where you log bids for each key competitor every time you bid against them. This takes only a few minutes and can be very useful.

Notice the *% above cost* column on the competitor bid forms. To calculate the figures in that column, first subtract your cost from their bid. Then divide the difference by your cost. Positive numbers show the percentage other contractors bid *above* your cost. Negative numbers show the percentage other contractors bid *below* your cost.

Notice that we don't know anything about their labor, material and subcontract costs. Instead, we're using our own estimated costs to make guesses about their bids.

How to gather information about your competition

Of course, market-based pricing is possible only if you know how much the competition bid on jobs already awarded. There are several ways to collect that information. When you bid government jobs, the winning bid is always disclosed. Usually the top three bids are made public. Sometimes all bids will be disclosed. It's harder to collect this information on private work.

I've found that owners are usually willing to answer questions about losing and winning bids after the contract is signed. Just call the customer after they've let the job. Ask them who got the job and for how much. Even on commercial jobs, most owners will volunteer something like, "Yeah, you guys were $23,000 too high on that one," or, "You were probably within 5 percent."

Some owners will avoid your questions. They'll say something like, "Well, I don't have the file in front of me right now," or, "All I'm going to say is that yours wasn't the low bid." In either case, you haven't learned a thing.

Here's a way that works even with the one-time prospects you'll never see again. The key is to make your request early, while the customer still needs your cooperation. Every time you get a request for a quote, try this:

Competitor Bid File

Company: **Notes:**

Location:

Principals:

Date	Description	Location	Their bid	Our cost	% above cost	Comments

Figure 12-1 Competitor bid form

Competitor Bid File						
Company: Able Homes			**Notes:** Smaller builder - aggressive in low to medium			
Location: Sarasota, FL			priced single family			
Principals: John Able						
Date	**Description**	**Location**	**Their bid**	**Our cost**	**% above cost**	**Comments**
Jan. 10	Single family	Sarasota	98,000	90,000	8.9	
Feb. 21	Single family	Sarasota	80,000	74,000	8.1	
Apr. 15	Single family	Sarasota	102,000	91,000	12.0	
May 23	Single family	Sarasota	100,000	95,000	5.3	
Oct. 23	Single family	Sarasota	75,000	68,000	10.3	

Figure 12-2 Bid file, Able Homes

Thank the caller for the request. Tell your prospect that you'll be glad to work up a very competitive figure. Then say, "I'd like to ask a favor. Once all your numbers are in and you've selected a contractor, please let me know how my price compared with other bidders."

Some people will be a little surprised by that. A builder who's just shopping and doesn't expect to award the job to anyone will be evasive or noncooperative, or both. In either case, it's nice to recognize a "shopper" in advance. Most serious prospects recognize that they need your help and will be anxious to cooperate. When people want something from you, they'll usually agree to a reasonable request. After all, you're doing the estimate for free and what you're asking costs them nothing.

Most people are as good as their word. If they agree with your request at the outset, they'll probably follow through when they've let the job.

Bid comparison doesn't have to be complex. Don't spend a week setting up a complicated computer spreadsheet for this. Make copies of the sample form or adapt it to suit your needs. Enter information from jobs bid last month and last year if you can. Start a file on each major competitor. List all their bid prices in order of oldest to most recent. When you've filled a few lines for each competitor, go on to the next step.

Competitor Bid FIle						
Company: Baker Builders			**Notes:** Commercial Builder - Bonds up to $ 500,000			
Location: Tampa, FL						
Principals: Bob Baker						
Date	**Description**	**Location**	**Their bid**	**Our cost**	**% above cost**	**Comments**
Jan. 14	Office	Sarasota	334,000	269,000	24.2	
Mar. 28	Office	Tampa	478,000	410,000	16.6	
May 16	Office	Sarasota	331,000	261,000	26.8	
Dec. 23	Office	Tampa	289,000	253,000	14.2	

Figure 12-3 Bid file, Baker Builders

Competitor Bid FIle						
Company: Charlie's Custom Homes			**Notes:** Custom Home Builders - $150,000 to $500,000			
Location: Sarasota, FL			bracket			
Pricipals: Charles Cobb						
Date	**Description**	**Location**	**Their bid**	**Our cost**	**% above cost**	**Comments**
Jan. 12	Single family	Sarasota	149,000	122,000	22.1	
Feb. 5	Single family	Sarasota	195,000	163,000	19.6	
Mar. 14	Single family	Sarasota	178,000	148,000	20.3	
Aug. 22	Single family	Sarasota	159,000	130,000	22.3	
Nov. 15	Single family	Sarasota	175,000	145,000	20.7	

Figure 12-4 Bid file, Charlie's Custom Homes

Competitor Bid FIle

Company: Donnels Construction **Notes:** Multi-Family Builder

Location: St. Petersburg, FL

Principals: James Watt

Date	Description	Location	Their bid	Our cost	% above cost	Comments
Jan. 28	Apartments	Tampa	979,000	845,000	15.9	
Feb. 14	Apartments	St. Pete	674,000	594,000	13.5	
Mar. 21	Apartments	Sarasota	903,000	773,000	16.8	
July 24	Apartments	Sarasota	1,778,000	1,520,000	17.0	
Dec. 1	Apartments	Sarasota	1,060,000	910,000	16.5	

Figure 12-5 Bid file, Donnels Construction

Competitor Bid FIle

Company: Ed's Construction **Notes:** Bids anything from $150,000 to $500,000

Location: Tampa, FL

Principals: Edward White

Date	Description	Location	Their bid	Our cost	% above cost	Comments
Feb. 20	Office	Naples	490,000	390,000	25.6	
Apr. 1	Single family	Tampa	199,000	202,000	(1.5)	
May 8	Office	Sarasota	620,000	515,000	20.4	
Aug. 15	Single family	St. Pete	149,000	148,000	.7	
Dec. 12	Office	Tampa	290,000	253,000	14.6	

Figure 12-6 Bid file, Ed's Construction

Competitor Bid File						
Company: Family Homes			**Notes:** Large builder of single-family homes			
Location: Sarasota, FL						
Principals: Fred Flintstone						
Date	**Description**	**Location**	**Their bid**	**Our cost**	**% above cost**	**Comments**
Jan. 10	Single family	Sarasota	101,000	90,000	12.2	
Jan. 12	Single family	Sarasota	139,000	122,000	13.9	
Feb. 7	Single family	Sarasota	112,000	97,000	15.5	
Mar. 14	Single family	Sarasota	166,000	148.000	12.2	
Apr. 15	Single family	Sarasota	104,000	91,000	14.3	
May 23	Single family	Sarasota	107,000	95,000	12.6	
Aug. 22	Single family	Sarasota	148,000	130,000	13.8	
Nov. 3	Single family	Tampa	180,000	162,000	11.1	
Nov. 15	Single family	Sarasota	171,000	145,000	17.9	

Figure 12-7 Bid file, Family Homes

Graph Your Competition

Figure 12-8 is a graph that shows how AEI Builders might use the information they've collected about their competition. AEI is bidding on a residential remodeling project. The estimated cost is $100,000. That includes construction costs and job overhead, but not company overhead or profit. Notice the line labeled "cost" drawn at 100 on the left scale. The line shows that we estimated our cost at $100,000.

The next line up from cost shows cost plus overhead. It's drawn at 108 on the left scale because our normal overhead is 8 percent of cost. This information comes from our business plan. So cost plus overhead is $108,000. The top line on this graph shows the normal selling price: cost plus 8 percent overhead and 8 percent profit margin, or $116,000. Without competitive analysis we would probably bid this job at $116,000. Is this the right market-based price?

The companies bidding against AEI Builders on this job are Able Homes, Charlie's Custom Homes, and Family Homes. Let's use our competitor bid file to anticipate the bids we can expect from these companies. Figure 12-2 lists what we know about Able Homes.

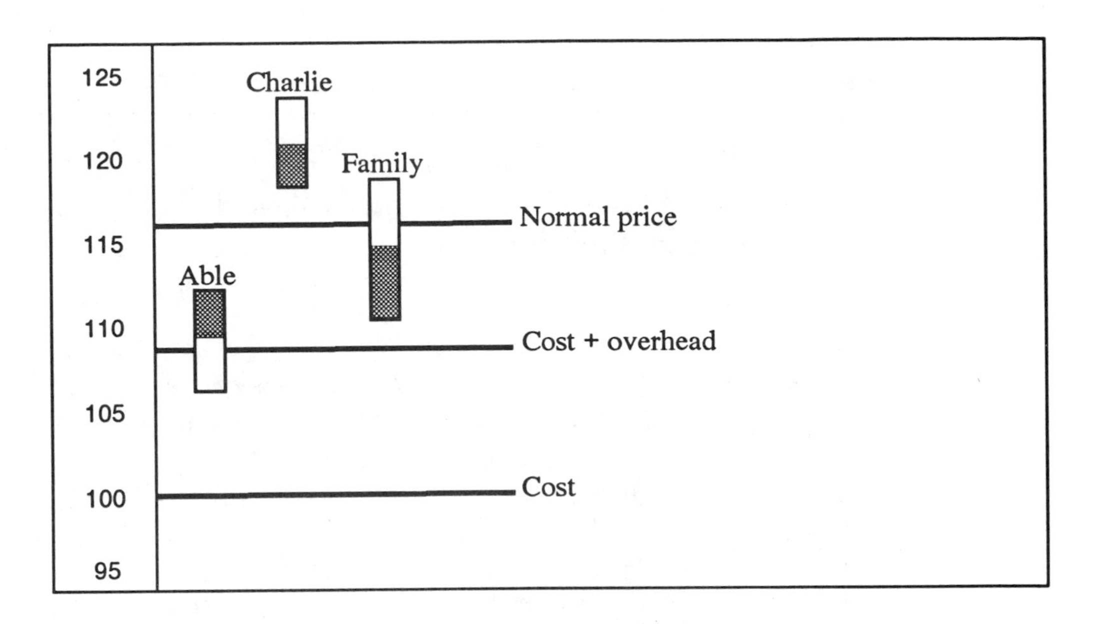

Figure 12-8 Bid file, Competitors' likely bids

Able has bid against AEI Builders five times. Their bids have ranged from 5.3 to 12 percent above AEI Builders' costs in the past. Notice the column labeled *% above cost.*

We know our cost is $100,000. If Able's price is between 5.3 and 12 percent higher than our cost on this job, their bid will range from $105,300 to $112,000. Notice that I've drawn a bar on Figure 12-8 to show the range of $105,300 to $112,000 for Able Homes. Remember, those points are based on AEI Builders' cost, not our bid price. We don't know what Able's markup rate is and we don't know what Able's costs are. We're just anticipating Able's bid price.

I'm sure you've noticed that AEI Builders has a hard time competing with Able. The bid prices Able submits are well below AEI's cost plus normal markup.

From Figures 12-4 and 12-7 I've plotted the range of bid prices for Charlie's Custom Homes and Family Homes. Looking at the price ranges plotted on Figure 12-8, you'll see right away that one competitor (Able) has always bid lower than AEI Builders. Another (Family) has bid in about the same range, and the third (Charlie) has always bid higher than AEI Builders.

Remember, this isn't a picture of what's sure to happen on this job. It's a picture of what has happened in the past. But this graph should help you decide what price to assign to this bid.

Let's Narrow It Down Some More

The bars in Figure 12-8 show the likely range of prices for each of the three competitors. (Don't worry about the shaded portion of those bars yet.) Now, let's limit that range somewhat. Are those companies likely to bid in the top end, the middle, or the lower end of their range on this job?

Here's where you apply knowledge about current market conditions, the job and the competition. How busy is Able? If he's swamped with work, he'll probably bid at the top of his range. If he's hungry, he'll be closer to the low end. How's the market generally? Is everyone busy, or is everyone scrambling to keep crews busy?

Look for a pattern in competitor's bids

Is there a pattern in the competitive bid record that might give you a clue? If Able's more recent jobs were in the higher range, you might expect Able to bid this job in the top third of his range. If you know Family needs work right now, you can expect them to bid in the low end of their range.

Suppose your competitive analysis is that Able will bid in the high end and both Charlie and Family Homes at the low end of their ranges. I've shown that on the graph by shading each of the three bars.

Now comes the subjective part of market analysis. Make some judgments based on what you know (or can guess) about the competition. Of course, no one can predict bids down to the last dime. But you can guess on a range of likely bids — and probably within the upper, middle or lower one-third of that range.

Now, think about what AEI Builders should make on this job and how much we should charge. What do you think Figure 12-8 is telling us?

How Will AEI Builders Handle This Bid?

Let's assume you're right on track for reaching the company goal of $6 million in contract volume this year. Assume also that there's nothing special about this job. You're not overloaded with work, but you don't need the job badly, either. Based on the information in this graph, what price would you quote on this job? Remember, your normal markup suggests a price of $116,000.

I've asked this question many times in seminars I've presented. Students usually suggest bidding about $112,000. What do you think?

We can see that two of the three competitors will probably bid well below $116,000. AEI is going to lose if they bid $116,000. No estimator likes to lose, of course. We like to win. We don't like getting shut out of work. Should AEI discount their price? Here's my view.

Estimating Isn't an Ego Trip

It isn't necessary to win every bid

Based on AEI Builders' current workload and budget, there's no reason to chase after this job. It's small and the profit will be slim. Don't try to satisfy your competitive instincts on every bid. You're paid to develop a workload that maximizes company profits, not to win every job. I know of a painting company whose estimator boasted that he won just about every job he bid on. Perhaps I should say I knew of a painting company. It's a fact of life that you're going to lose more jobs than you win. Most estimators bid ten jobs for every one they get. My advice is to forget about trying to win all the time. Instead, submit prices based on good business judgment.

"Most estimators bid ten jobs for every one they get."

My business judgment tells me that this isn't a job for AEI Builders. When Able's going to bid, I shouldn't even bother estimating. On a $100,000 job, I'll pass when the range of markups probably doesn't even cover my overhead. Now, if this was a million dollar job, I'd be more interested. But this is a small job for AEI Builders. Why take a small job at a price that'll barely cover overhead?

If your first reaction was to cut the price on this job, don't be embarrassed. You're not unusual. Most people would say, "We've got to cut the price here, because this chart says we can't get the job at our regular markup." But there's a better way to handle this.

Know when to raise your bid

Raise your bid to $120,000.

This tactic will do two things for you.

1) You'll shut yourself out for sure on a job you really don't want or need.

2) You'll throw your competition a smoke screen.

There's only a slim chance you'll get this job anyway, even if you cut the bid price to below normal markup. Remember also that actual labor costs have been running about 10 percent over estimated labor costs on residential work. There's more risk here and very little margin for error. That smells like trouble to me.

There's another reason to highball this job. If your competition is sharp, they're doing the same analysis I've recommended here. They're plotting your range of bids. Let's throw them a curve. They'll see your bid price, and say, "Why did AEI come in so high? They must be loaded with work. Or maybe they've increased their markup. Maybe we should fatten our markup a little, too."

This strategy should pay off on the next job when you come up against the same bidders. But this time you may need the job as filler between larger projects. Your competitors may think, "AEI has been bidding high lately. We don't have to worry about them. We can bid a little higher on this one." That should give you an opening and a few extra dollars when the job is done.

Don't be the low bidder all the time

There's an important point here: Don't try to be low bidder all the time. You'll wind up with a heavy workload of low profit or zero profit jobs. I know it's a jungle out there. But some jobs are better left to the competition. Resist the instinct to go for the action. Instead say, "I just don't want this job at the price it's going for. I know what the winning bid will be and I don't want it at that price. It doesn't fit my long-range plan."

Let's Wrap This Up

If you don't compete in the open bid market, this chapter may not apply to your company. But most of us have plenty of competition. And if you do, keep track of the competition. Keep good records. Make those follow-up phone calls on bids you've submitted. Know where the competition is headed so you know where the opportunities are.

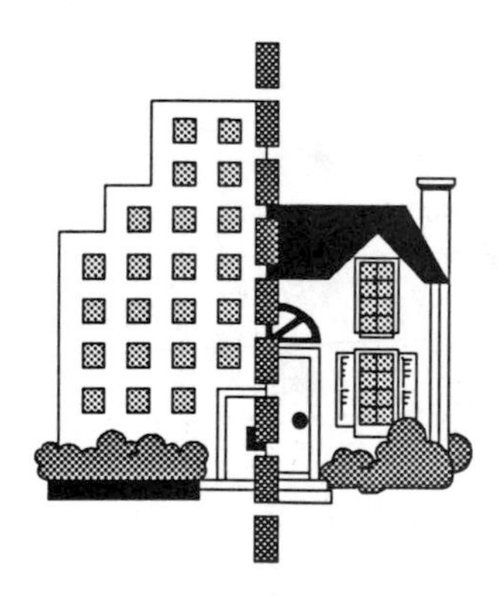

While you're at it, ask yourself some important questions. Does what you're doing today provide enough income? Or should you make some changes now, before the work you're chasing dries up entirely? The time to change your product is before you have to. While you're making money at what you're doing today, take a job at something you haven't built before. Look for areas and types of work that show promise for the future. You probably won't make any money the first time you try something new. Learning and experience has to come before profit. But you can afford to do that as long as you're getting by on other work. The best time to experiment is when everything is going great and it seems there's no reason to experiment at all.

That brings us to the close of both this chapter and the book. In some ways, this is an unconventional estimating reference manual. The range of information covered here goes all the way from the most basic to much more advanced topics you've probably never even considered before. I hope I've opened your eyes to better ways to estimate and better ways to meet the competition. Most of all, I hope that what I've explained helps you make a better living as a skilled professional construction cost estimator.

Index

A

"A" sheets 32
Abbreviations, in estimates 49
Acceptable cost overrun 235
Acceptance form 121
Accounting programs 153
Accounts receivable programs . . .153
Accuracy, estimating 25, 235
Active bids notebook 37
Actual job costs . . .132, 177, 187-188
Actual overhead 192
Actual owner's salary 206
Adapting to market conditions 23
Add for Overhead & Profit dialog box 165
Adding to costbook, National Estimator 165
Adding to estimate, National Estimator
  cost lines 164-165
  description 164
Adjusting
  bids 19
  manhours 184
  markup 248
  prices 248-249
Adjustment for risk 229
Advertising budget, tracking 198
Allowance
  contingency 46, 114
  rebar waste 142-143
  waste 67
Allyn, Marques 155
Analysis, cost overruns 238
Analyzing
  competition 19, 253
  plans 33-36
  risk 229
Annual budget 191, 196
  exercise 199
  reason for 208
Annual business plan 195
Annual sales volume, goal 202
Anticipating competitor bids 262
Applying markup 216
Architects, dealing with 34
Architectural sheets 32
Arithmetic errors 49
Asset utilization 229, 240
  effect on markup 242
  effect on profit 251
  exercise 243, 250
  net effect 251
Assets
  company 16
  effect on markup 241
Audit trail 131, 143
Auditing expense 207
Auto-Mate for Builders 153
Average
  markup 220
  markup, adjusting 229
  profit, industry 209

B

Backcharge, equipment 208
Backfill take-off 144-145
Bad weather 230
Basement plan 33
Basic Building Data 58, 132
Basic
  budget calculations 209
  management salary 205
Batterboards 146
BBD .58, 132
Beginning an estimate, National Estimator 158
Bench marks 134
Best profit 215
Bid
  adjusting 19
  adjustment 229
  files 257-262
  filing 37
  history .218
  history, planning without 227
  package .32
  predictability 254
  preparing 117
  requesting from subs 57
  reviewing 78
  scope, limiting 90
  shopping 72
Bidding
  in recession 20
  procedure flowchart 13
  sequence 33
Bidding forms on CD Estimator . .155
Bill of materials 96
  cross-referencing 107
Billing rate, equipment 207
Bills, filling 188
Blinking cursor 158
Block 43
Blockouts 143
Board feet 103
  calculating 43
Bond, performance 73
Bonding
  capacity 212
  limit 202
Bonuses .206
Bookkeeping programs 153
Bookmarks, Web browser 176
Bottom line
  preliminary 208
  tracking 198
Breakdown
  estimate analysis 240
  in bid .117
Break-even point 226
Brick 43
Bridge repair 24
Budget
  annual 191, 196
  calculations, basis 209
Building laborer, wage rate 163
Building permit 46
Business
  cycles 20
  direction, changing 264
  multiple 27
Business plan 17, 191
  and loan application 22
  changing 27
  long-term 21
  markup 19, 229
  recommendations 210-211
  time period 22
Business planning questions 22
Buttons on tool bar 158

C

"C" sheets 32
Cabinet demolition 104
Cabinets, estimate detail 108
CAD programs 153
Calculating
  cost overrun 234
  overhead 191
  projected overhead 225
  quantities 141
Capacity, bonding 212
Career in estimating 8, 12
Cash
  as company assets 240
  under-use 246
Cash flow, effect on markup 242
Categories
  comparing jobs by 230
  cost 181
  estimate 150
  estimate breakdown 240
CD Estimator 153
  databases included 154, 167
  installation 156-157
  setup 155-156
Changes
  business plan 27
  operating 210-211
  operations 191
  overhead 193
Changes, National Estimator
  estimate costs 163
  estimate descriptions 164
Chargeback, equipment 207
Charging for
  overhead 18
  profit 18
  supervision 114
Charting overhead 225
Checking
  estimates 47, 117
  estimating accuracy 235
Checklist
  estimating 79
  job site .55
  plant and equipment 85
  project overhead 77
Chief estimator 11
Chimney, bidding 35
Choosing projects 17, 19, 53
Classification
  estimating 39
  estimator 11
  job costs 181
Classifying labor costs 182
Clearing the site 41
Clerical salaries 207
Closed bids 220
Closing the sale 120
Collecting competitor bids 255
Color key, National Estimator 159
Column heads, National Estimator 157-158, 161
Column lines, plans 136
COM port for modem 169-170
Comment column, on time card . .187
Commercial estimate summary .81-84
Commercial estimates 126
Communications program, Excalibur 155, 168-169
Compaction 145
Company
  assets 16
  facilities, utilization 247
  overhead 46, 191
  profile 195
Comparing
  competitor bids 257
  costs over time 197
  estimates 149
Competition
  analyzing 19, 253
  graphing 260
  identifying 53
  learning about 255
Competitor Bid Form 256
Competitor bids
  collecting 255
  pattern in 262
Completed project worksheet 230, 232
Completion and Acceptance Form 120, 121
Compound
  profit increase 208
  sales increase 202
Compressed files 174
Computer estimating 153-175
  CD Estimator 153-154
  National Estimator, using . .157-167
Computer modem, using 168-170
Computer requirements, National Estimator 155
Computerized cost records 180
Computing markup 87
Concrete 42
  plan 133
  quantity 141
  specifications 137
  take-off 135
Conditions, working 38
Confusing your competition 263
Construction
  cost estimating by computer 157-167
  costs, forecasting 203
  drawings 29
  industry statistics 6, 209
Construction Estimating program group 154, 157, 169
Construction management programs 153
Construction Specifications Institute 39, 181
Contingency 46, 114, 152
Contract, estimates as 117
Contractor's Bulletin Board System
  (BBS) 155, 168-175
  dialing .171
  downloading files 172-174
  E-mail messages 175
  exit, how to 174
  exploring 175
  libraries 172
  logging on 172, 174-175
  on-line assistance 171
  sorting file lists 173
  sound card compatibility 175
  System Operator 175
  technical support 154, 172
  tool bars 172
  uploading files 172
  welcome screen 171
  .ZIP files 173
Contractor's BBS and Contractor's Home Page 168
Contractor's Home Page, Internet 168, 175-176
Control points 134
Controlling sales volume 201
Conversion
  inches to feet 67
  markup to profit 223
Copying errors 49
Copying in National Estimator
  cost lines 162-163
  descriptions 162
  partial lines 162
Cost
  analysis 180
  breakdown, in bid 117
  element, risk by 236
  guides .38
Cost data
  file 36
  using .187
Cost lines
  adding to estimate 164-165
  color key, National Estimator . . .162
Cost of estimating 53

Cost overruns .230
- eliminating .239
- exercise .234
- reasons for .237

Cost per manhour .146
Cost recording .177
Cost records .132, 149, 178
Cost updates, National Estimator .155, 173
Costbook lines, National Estimator
- adding to .165
- copying .162-163
- editing .162
- inserting .161

Costbook page, selecting .159
Costbook updates, downloading .173-174
Costbooks, National Estimator
- copying to hard disk .156
- databases .154, 158, 167
- opening .154, 158
- revisions .173

Costs
- changing in National Estimator .163
- finding in National Estimator .159
- forecasting .203
- monitoring .197
- owner's .205
- unexpected .36
- unit .61
- verifying .78

Countertops, estimate detail .111
Covering overhead .246
Craft@Hours column .158
Craftsman Support Forum .173, 174-175
- Miscellaneous Utilities .174
- Upgrades & Updates .173

Craftsman technical support .154, 175
Crew
- identifying .146
- productivity .61
- using your own .239
- wage rate .105

Critical path .68
Cross-referencing, bill of materials 107
Crossfoot .106, 150
CSI Masterformat number 39, 181, 189
Curbs, take-off .144
Current costbook .161, 167
Current estimate .161
Current jobs, tracking costs .182
Cursor, National Estimator .158
Custom installation, CD Estimator 156
Custom markup .18
Customizing your estimate .164-165
Customer relations .31, 95
Customer's
- copy, Detail Estimate .117
- responsibility .95

Cutting prices .249
Cycles, business .20

D

Daily time slip .184
Databases, CD Estimator .154, 167
- updating .168

Decision chain .13-16
Default costbook, National Estimator .167
Demographics .23
Demolition .91
- cabinet .104
- estimate .102
- estimate detail form .98
- materials .102

Deposits, insurance and tax .45
Descriptions, copying, National Estimator .161
Destination directory .156
Detail drawings .31
Detail, in bid .117
Detailed
- cost estimate .53, 60
- estimate .7
- estimating steps .51

Diagnostics for modem, Windows .95, 169
Dialing Contractor's BBS .171
Dialing Directory dialog box .171
Differences, estimate types .125
Difficulty, job .236
Dimensions
- re-using .132
- wall .33

Direct billing, equipment .207
Directory
- destination .156
- installing CD Estimator .156

Disk space
- required for CD Estimator .156
- saving .155
- Watch Me .155

Disputes, avoiding .90, 118
Documents, bid .32
Doors, on plans .34
Doubling sales volume .202
Dowels .142
Downloading files, Contractor's BBS .172-175
Drafting programs .153
Draw, owner's .205
Drawings
- footings and foundations .34
- marking .132
- perspective .31

Dual profit .37
Duplicate paste, undoing .162

E

E-mail, Contractor's BBS .173
Early price quotes, avoiding .31
Economic cycles .20
Edit Directory Entry dialog box .171
Effect on profit, asset utilization, .251
Efficiency, supervisors .240
Electrical work .44
Electricity .46
Electronic index, National Estimator. .154, 159
Elevation drawings .31, 35, 96
Employee time records .183
Employees, training .212
Enter Keyword to Locate box 159-160
EPA funding .25
Equipment
- as company assets .240
- computer .155
- cost .106
- direct billing .207
- expense .207
- project overhead .76
- under-use .244

Errors, math .62
Escalation .46
Estimate
- analysis .240
- breakdown, grouping .240
- checking .47, 117
- customer's copy .117
- detailed .7
- filing .37
- log .99
- program, National Estimator 156-167
- requesting from subs .57
- reviewing .78
- summary .79
- take-off .38, 58

Estimate Detail Form .89-107, 131, 146-147
- cabinets .108
- concrete .147
- countertops .111
- demolition .98
- flooring .113
- sink .112

Estimate name, National Estimator 166
Estimate, beginning .158
Estimate, changing, National Estimator
- costs .163
- descriptions .164

Estimate summary .79, 118, 131, 150-151
- industrial and commercial .81-84
- residential .86

Estimate Summary Form .115
Estimate Take-off Sheet .125, 130
Estimate window .158
Estimate1 .157
Estimating
- as a career .8, 12
- checklist .79
- classification .39
- databases .154, 167
- forms, identifying .133
- mistakes, reducing .47
- overhead .225
- program, National Estimator 156-167
- reference .189
- rules .91
- sequence .33
- skills .7
- steps .51
- success rate .12

Estimating forms on CD Estimator 155
Estimator
- classifications .11
- profile .10
- qualifications .9
- training .212

Evaluating projects .53
Events, sequence of .69
EXCAL.EXE .175
Excalibur communications program .155, 169-175
- Client icon .169
- Client Update dialog box .175
- dialog boxes .169-171, 175
- installing .169
- License Agreement dialog box .168
- new versions .174

Excavation
- costs .42
- footings .136
- quantity .141
- take-off .135

Exclusions
- bid .47, 90
- scope of work .146

Exercises
- annual budget .199
- asset utilization .243, 250
- cost overrun .234

Extending costs .62, 103
- with National Estimator .154

Extensions, price .49

F

F.I.C.A. .45
F.U.T.A. .45
Facilities, utilization .247
Factor, adjustment, labor .185
Fifteen estimating steps .51, 128
File menu, National Estimator .166
File Name box .166
File save as .166
Files tool bar, Contractor's BBS .172
Filing estimates .37
Filling out forms .133
Finding subcontractors .56
Finishes .44
Fire insurance .46
Fireplaces, bidding .35
Fitting jobs to long-range plan .264
Fixed fee .216
Flatwork, concrete .42
Floor plans .31
Flooring, estimate detail .113
Flowchart, bidding procedure .13
Footings
- excavation .136
- finding on drawings .34
- layout .136
- take-off .138

Forecasting
- construction costs .203
- overhead .204

FORMS directory, CD Estimator .155
Formwork take-off .67, 143-144
Foundation
- layout .42, 135, 148
- plan .33

Fractions, errors in calculating .49
Fringe benefits, owner's .205
Full installation, CD Estimator .156

G

General condition items .7
Goals
- annual sales volume .202
- business .22
- productivity .248

Grades, rebar .142
Graph, markup vs. volume .221
Graphing competition .260
Gross profit .203, 215
Gross sales, predicting .201
Guide to National Estimator
- printing .155
- video .155

Guidelines, markup .218
Guides, cost .38

H

Hard disk, saving to .166
Haunches, concrete formwork .143
Hazardous work .230
Heading for estimate .158
Help, National Estimator 155, 167-168
Hidden costs, overhead .194
History, bid .218
Home Page, Contractor's, Internet .168, 175-176

I

Icons
- costbooks .154
- Excalibur .168
- National Estimator .154-155

Incentive pay .206
Inches, converting .63, 67, 139
Income statement .196
Increasing sales .201, 204
Index, National Estimator .156-160
INDEX.TXT .155
Industrial estimate summary .81-84
Industry, construction, average profit .209
Inserting a line, National Estimator 161
Install Icons? dialog box .169
Installation
- CD Estimator .156-157
- Excalibur .169
- National Estimator .156-157

Insulation .44
Insurance .44
Internet, Contractor's Home Page .168, 175-176
Internet, using for National Estimator .168

J

Job costs, recording 36-37, 132
Job identification 133
Job site
checklist 55
obstructions 55
visit 36, 55, 129
Job size, effect on markup 241
Job specifications . .29, 120, 122, 137
Job type, risk by 237
Journeyman estimator 11
Junior estimator 11

K

Keyword search, National Estimator 159-160

L

Labor
burden 44-45, 206
costs, classifying 182
costs, extending 149
force, as company asset 240
non-productive 116
project overhead 76
Laps, rebar 142
Last-minute bidding 72
Layout, foundation 42
Learning about competition 255
"Leaving money on the table" . . .253
Legal expense 207
Legibility, importance of 50
Liability insurance 45
Library tool bar, Contractor's BBS 172
Limits, annual sales volume 202
Loan application, business plan and 22
Log, estimate 99
Logging on, Contractor's BBS 172, 174-175
Long-range plan 21, 191, 211
fitting jobs to 264
Lotus 1-2-3 forms 155
Lowering prices 19
Lumber, pricing 43, 103

M

Main menu, Contractor's BBS . . .172
Man-day .148
Management
personnel, under-use 245
salary 204
Manhour
errors 49
estimates 38, 104, 148
Manhours 106
on bid .118
recording 183
Margin 196, 215
Market
conditions 20, 26, 262
conditions, adapting to 23
effect on markup 241
switching 21
Market-based pricing 229, 253
Markup
adjusting 19, 248
asset effect on 241
average 220
computing 87, 114
custom 18
effects on 241-242
market conditions, effect on 25
normal 17, 216
percentages, National Estimator .165
Markup vs. volume 218, 220
Masonry 43
Masterformat number 39, 189
Material
costs 61, 148
prices, trends 47
schedules 33
Materials
demolition 102
foundation layout 148
Maximum profit 224
MBF (thousand board feet) 103
Measurements, errors in 49
Medicare tax 45
Menu bar, National Estimator 157, 160
Metalwork 43
Microsoft Excel forms 155
Microsoft Video for Windows 157
Microsoft Word forms 155
Microsoft Works forms 155
Miscellaneous expense 207
Mistakes in estimates 47, 49, 62
Modem
setting dialog box 170
Setup and Defaults dialog box . .170
testing 168-169
using for cost updates 168
Moisture protection 44
Molding 43
Monitoring productivity 248
Mouse pointer 158
Mouse, using for National Estimator 157, 159
Multiple businesses 27, 239

N

Naming the estimate, National Estimator 166
National Construction Estimator 154, 159, 163, 167
National Electrical Estimator 154, 167
National Estimator program
changing your estimate 163
changing the costbook 163, 165
cost updates 155, 168
costs, adding 164
databases 154, 167
electronic index 154
extending prices 154
guide to, video 155
Help 155, 167-168
icons 154-155
installation 154-155
markup, adding 165
open costbook 154
program updates 173, 176
split screen 154
starting 154
tax, adding 165
totaling columns 154
user costbooks 174
using 157
wage rates 163
Watch Me 155
Windows 154
National Painting Cost Estimator 154, 167
National Plumbing & HVAC Estimator 154, 167
National Renovation & Insurance Repair Estimator 154, 167
National Repair & Remodeling Estimator 154, 167
Negotiated contracts 254
Net
effect, asset utilization 251
profit 215
profit, graph 226
Netscape .175
New construction costs 78
96CONST.CBK 154
96ELECT.CBK 154
96INSUR.CBK 154
96PAINT.CBK 154
96PLUMB.CBK 154
96REP.CBK 154
Non-productive labor 116, 185
Non-responsive bid 87
Normal
margin .196
markup 17, 227
markup, adjusting 229
Notebook, active bids 37
Numbering, estimates 99

O

Obstructions, job site 55
Off-site labor 116
Office
expense 207
temporary 46
Omissions .49
On-line assistance, Contractor's BBS 175
One-year plan 191
Open bids 220
Open Costbook, National Estimator 154, 167
Operations, changes in 191
Optional bids 110
Order of take-off 133
Orion Designs 153
Overhead 114, 215
adding to estimate, National Estimator 165
and non-productive time 185
calculated 224
calculating 191
changes 193
charging for 18
checklist 77
covering 246
effect of job size 152
forecasting 204
project 7, 46, 76
recap form 193
Overruns .230
eliminating 239
reasons for 237
Owner's
costs 205
fringe benefits 205
salary 205
Ownership, businesses 205

P

Page number box 159
Password box, Excalibur 171
Pasting a line, National Estimator .161
Pattern
in competitor bids 262
of errors 238
Payroll accounting 183
Performance bond 73
Personnel, under-use 245
Perspective drawings 31
PKUNZIP 174
PKWARE 174
Plan
business .17
long-term 21
package .32
view 96
Plans
analyzing 33-36
architectural 29
marking 132
openings on 34
Plant and equipment checklist 85
Plumbing .44
Population distribution 23
Pre-assembly 71
Predictability, bids 254
Predicting sales volume 201
Preliminary
bottom line 208
quote, avoiding 30, 95
Preparing a budget 199
Preview estimate, National Estimator 166
Prices
adjusting 248-249
lowering 19
quoting .95
raising .249
Pricing
strategies 253
trends 47
Print estimate, National Estimator .166
Print Preview, National Estimator .166
Printer for CD Estimator 155
Printer options, National Estimator 166
Probable range, competitor bids . .261
Problems with subcontractors 74
Production assets 240
Productive labor 116
Productivity
crew 61, 236
estimates 38
goals 248
worker .186
Profile
estimator 10
sample company 195
Profit 215
adding to estimate, National Estimator 165
before-tax 196
charging for 18, 79
compound increase 208
curve 222
dual 37
increasing 208
industry averages 209
job size, effect on 152
on remodeling estimate 114
reviewing 27
utilization effect on 251
Profit margin
reductions in 204
tracking 198
Program group, Construction Estimating 154, 157, 169
Program updates, National Estimator 173, 176
Project
acceleration 71
adjustments 248
choosing 17, 19
evaluating 53
manager 11, 178
mix 236
risk 19
schedule 68
summary 230-231
worksheet form 233
Project manager, training 212
Project overhead 7, 46, 191
checklist 77
equipment 76
labor 76
supporting costs 76
Proposal letter 118
Published cost guides 38, 132

Q

Qualifications, estimator 9
Quantity take-off 38
Questions
business planning 22
site visit 36

Quota
- estimate as .117
- worker's .186

Quoting price .95

R

Rain delay .185
Raise bid, when to .263
Raising prices .249
Range
- competitor bids .261
- correct .200
- estimate accuracy .25

Rates, Workers' Comp .45
Ratio, jobs won .254
Read Me, National Estimator 155, 157
Reasons for cost overruns .237
Rebar .142
- specifications .137
- take-off .68, 135

Recession, bidding in .20
Recommendations
- business .210-211
- subcontractors .56

Recording job costs .36-37
Reductions, profit margin .204
Reference
- estimating .189
- numbers .107

Reinforcing bars .142
Relationship, markup to volume .222
Relative worth sales .246
Remodeling .24, 89
- costs .78
- estimate sample .93

Renovation .24
Repair work .89
Repairs, adjoining property .46
Requesting bids from subs .57
Residential estimate summary .86
Results, monitoring .197
Review estimate .78
- profits .27
- scope of work .96

Revised cost files, downloading .173
Rewards, estimating career .12
Risk .19
- analysis .229
- by cost element .236
- by job type .237
- estimating categories .217

Roofing .44
Rules, estimating .91

S

Salary
- management .204
- owner's .205
- senior estimator .206

Sales
- increase, compound .202
- relative worth of .246
- tracking .198
- volume, predicting .201

Sales tax, adding to estimate .165
Sample
- remodeling estimate .93
- take-off .63

Saving the estimate, National Estimator .166
Savings, overhead .194
Schedule, project .68
Schedules
- materials .33
- windows and doors .34

Scheduling, work crew .107
Scope of work
- defined for sub .73
- exclusions from .134, 146
- review .96
- subcontractor's .75

Scope, bid .90
Section drawings .31
Security, job site .55, 230
Select Destination Directory dialog box .169, 175
Select Program Manager Group dialog box .169
Senior estimator salary .206
Sequence
- plan reading .33
- project .69

Setting wage rates, National Estimator .163
Setup, CD Estimator .155-156
Sewer connection fee .46
Shareware programs .168
Shop labor .116
Shopping bids .72-73
Shortcomings, career in estimating .12
Sidewalk permit .46
Sink, estimate detail .112
Site clearing .41
Site visit .36, 55
- checklist .55

Skills, estimating .7
Slab estimate .146
Slide bar, page number .159
Smart bidding .229
Social Security tax .45
Soil, moving .42
Solutions
- annual budget .199
- asset utilization .244
- cost overrun exercise .235

Sound card compatibility .175
Sound system for Watch Me .155
Space on hard disk, saving .155
Spec building .21
Specialty contracting .24
Specifications
- concrete .137
- conflicts .54
- described .53
- job .29
- remodeling .122

Split screen, National Estimator .154, 160-161
Square foot costs .78
Square footings .140
Staff limits on sales .203
Stairs .34, 43
Starting National Estimator .154
Statistics, construction industry .6
Status bar, National Estimator .158
Steel take-off .68
Steps, fifteen estimating .51
Strategic pricing .253
Stringers .143-144
Subcontractors
- bids .71-75
- costs .106
- drawings for .33
- finding .56
- problems with .74
- reasons for using .239
- recommendations .56

Success rate, estimating .12
Summary estimate .79
Supervision .76
- charging for .114

Supervisors
- as company assets .240
- efficiency of .240
- time cards .183
- training .212

Supplies, availability near job site .55
Supply and demand .25
Supporting costs, project overhead .76
Surety bonds .46
Surprises, avoiding .91
Survey markers .134
Switch windows, National Estimator .161
Switching markets .21
Symbol, control point .134
System operator, Contractor's BBS .175

T

Take-offs
- estimate .38, 58
- footings .138
- how to .132
- sample .63
- sheet .129, 131
- software .155

Tax, adding to estimate .165
Taxes, payroll .45
Technical support, Contractor's BBS .154, 175
Telephone charges .46
Temporary office .46
Text lines, adding to estimate .164-165
Text, changing, National Estimator 164
Text/cost lines, color .159
Thermal protection .44
Tie wire .143
Time cards
- comment column .187
- filing .188
- supervisor's .183

Time period, business plan .22
Time records, employee .183
Time, wasted .186
Timekeeper .46
Title bar, National Estimator .157
TNE.EXE .168
Toilets .46
Tool bars
- Contractor's BBS .172
- National Estimator .157-158

Totaling columns, National Estimator .154
Tracking
- current job costs .182
- periods, budget .198

Traffic congestion .230
Training
- employees .212
- supervisors .212

Trends, pricing .47
Trim .43

U

Under-use
- cash .246
- equipment .244
- personnel .245

Undo command, National Estimator .162
Unemployment insurance .44
Unexpected
- costs .36
- events, recording .187

Unit costs
- applying .61
- defined .7
- in cost recording .179
- in published references .78
- on concrete estimate .146
- on remodeling estimate .103

Units of work .183
Unzipping a compressed file .174
Updates, National Estimator costbooks .168, 173-174
Upgrades & Updates, Contractor's BBS .173
User costbooks, National Estimator174
User's guide to National Estimator 155
User Information dialog box, Excalibur .170
Using National Estimator .157
Utilities .46
Utilities, menu bar, National Estimator .167
Utilization
- assets .229
- effect on profit .251
- of company facilities .247

V

Varying business plan price .19
Verifying costs .78
Video for Windows .157
Video guide to National Estimator 155
Visiting the job site .129
Volume
- concrete .42
- sales .201
- sales, increase .204

W

Wage rates .105, 149
- in National Estimator .163

Wall dimensions, checking on plan .33
Waste
- allowance .67
- rebar .142

Wasted time .186
Watch Me video .155, 156-157
Watchman .46
Water connection fee .46
WAV sound tracks .175
Weather, effect on bid .230
Web browser .175
Welcome screen, Contractor's BBS .171
Welcome To Excalibur! dialog box .168
Windows 3.1 and 3.11 .154, 156, 168
Windows 95
- modem testing .168-169
- using for CD Estimator .156, 168

Windows and doors, on plans .34
Wiring .44
Word processing programs .153
WordPerfect forms .155
Work
- crew, scheduling .107
- description, time cards .183
- units .183

Worker productivity .186
Workers' Comp .45, 230
Working conditions .38
Workload, suitable .251
Worksheet, completed projects .230
Worth, sales dollars .246
Written proposal .118
Yearly business plan .195

Z

ZIP files, Contractor's BBS .174

Other Practical References

Audiotapes: Estimating Electrical Work

Listen to Trade Service's two-day seminar and study electrical estimating at your own speed for a fraction of the cost of attending the actual seminar. You'll learn what to expect from specifications, how to adjust labor units from a price book to your job, how to make an accurate take-off from the plans, and how to spot hidden costs that other estimators may miss. *Includes six 30-minute tapes, a workbook that includes price sheets, specification sheet, bid summary, estimate recap sheet, blueprints used in the actual seminar, and blank forms for your own use.* **$65.00**

Basic Plumbing with Illustrations, Revised

This completely-revised edition brings this comprehensive manual fully up-to-date with all the latest plumbing codes. It is the journeyman's and apprentice's guide to installing plumbing, piping, and fixtures in residential and light commercial buildings: how to select the right materials, lay out the job and do professional-quality plumbing work, use essential tools and materials, make repairs, maintain plumbing systems, install fixtures, and add to existing systems. Includes extensive study questions at the end of each chapter, and a section with all the correct answers. **384 pages, 8½ x 11, $33.00**

Construction Estimating Reference Data

Provides the 300 most useful manhour tables for practically every item of construction. Labor requirements are listed for sitework, concrete work, masonry, steel, carpentry, thermal and moisture protection, door and windows, finishes, mechanical and electrical. Each section details the work being estimated and gives appropriate crew size and equipment needed. Includes an electronic version of the book on computer disk with a stand-alone *Windows* estimating program **FREE** on a 3½" high-density (1.44 Mb) disk. If you need 5¼" high-density disks add $10 extra. **432 pages, 11 x 8½, $39.50**

Residential Steel Framing Guide

Steel is stronger and lighter than wood — straight walls are guaranteed — steel framing will not wrap, shrink, split, swell, bow, or rot. Here you'll find full page schematics and details that show how steel is connected in just about all residential framing work. You won't find lengthy explanations here on how to run your business, or even how to do the work. What you will find are over 150 easy-to-ready full-page details on how to construct steel-framed floors, roofs, interior and exterior walls, bridging, blocking, and reinforcing for all residential construction. Also includes recommended fasteners and their applications, and fastening schedules for attaching every type of steel framing member to steel as well as wood. **170 pages, 8½ x 11, $38.80**

How to Succeed With Your Own Construction Business

Everything you need to start your own construction business: setting up the paperwork, finding the work, advertising, using contracts, dealing with lenders, estimating, scheduling, finding and keeping good employees, keeping the books, and coping with success. If you're considering starting your own construction business, all the knowledge, tips, and blank forms you need are here. **336 pages, 8½ x 11, $24.25**

Commercial Electrical Wiring

Make the transition from residential to commercial electrical work. Here are wiring methods, spec reading tips, load calculations and everything you need for making the transition to commercial work: commercial construction documents, load calculations, electric services, transformers, overcurrent protection, wiring methods, raceway, boxes and fittings, wiring devices, conductors, electric motors, relays and motor controllers, special occupancies, and safety requirements. This book is written to help any electrician break into the lucrative field of commercial electrical work. **320 pages, 8½ x 11, $27.50**

Estimating Plumbing Costs

Offers a basic procedure for estimating materials, labor, and direct and indirect costs for residential and commercial plumbing jobs. Explains how to read and understand plot plans, design drainage, waste, and vent systems, meet code requirements, and make an accurate take-off for materials and labor. Includes sample cost sheets, manhour production tables, complete illustrations, and all the practical information you need. **224 pages, 8½ x 11, $22.50**

Contractor's Year-Round Tax Guide Revised

How to set up and run your construction business to minimize taxes: corporate tax strategy and how to use it to your advantage, and what you should be aware of in contracts with others. Covers tax shelters for builders, write-offs and investments that will reduce your taxes, accounting methods that are best for contractors, and what the I.R.S. allows and what it often questions. **192 pages, 8½ x 11, $26.50**

Drafting House Plans

Here you'll find step-by-step instructions for drawing a complete set of home plans for a one-story house, an addition to an existing house, or a remodeling project. This book shows how to visualize spatial relationsnips, use architectural scales and symbols, sketch preliminary drawings, develop detailed floor plans and exterior elevations, and prepare a final plot plan. It even includes code-approved joist and rafter spans and how to make sure that drawings meet code requirements. **192 pages, 8½ x 11, $27.50**

Cost Records for Construction Estimating

How to organize and use cost information from jobs just completed to make more accurate estimates in the future. Explains how to keep the records you need to track costs for sitework, footings, foundations, framing, interior finish, siding and trim, masonry, and subcontract expense. Provides sample forms. **208 pages, 8½ x 11, $15.75**

National Building Cost Manual

Square foot costs for residential, commercial, industrial, and farm buildings. Quickly work up a reliable budget estimate based on actual materials and design features, area, shape, wall height, number of floors, and support requirements. Includes all the important variables that can make any building unique from a cost standpoint.
240 pages, 8½ x 11, $23.00. Revised annually

Basic Engineering for Builders

If you've ever been stumped by an engineering problem on the job, yet wanted to avoid the expense of hiring a qualified engineer, you should have this book. Here you'll find engineering principles explained in non-technical language and practical methods for applying them on the job. With the help of this book you'll be able to understand engineering functions in the plans and how to meet the requirements, how to get permits issued without the help of an engineer, and anticipate requirements for concrete, steel, wood and masonry. See why you sometimes have to hire an engineer and what you can undertake yourself: surveying, concrete, lumber loads and stresses, steel, masonry, plumbing, and HVAC systems. This book is designed to help the builder save money by understanding engineering principles that you can incorporate into the jobs you bid. **400 pages, 8½ x 11, $34.00**

Contractor's Guide to the Building Code Revised

This completely revised edition explains in plain English exactly what the Uniform Building Code requires. Based on the newly-expanded 1994 code, it explains many of the changes made. Also covers the Uniform Mechanical Code and the Uniform Plumbing Code. Shows how to design and construct residential and light commercial buildings that'll pass inspection the first time. Suggests how to work with an inspector to minimize construction costs, what common building shortcuts are likely to be cited, and where exceptions are granted.
240 pages, 8½ x 11, $39.00

Estimating Electrical Construction

Like taking a class in how to estimate materials and labor for residential and commercial electrical construction. Written by an A.S.P.E. National Estimator of the Year, it teaches you how to use labor units, the plan take-off, and the bid summary to make an accurate estimate, how to deal with suppliers, use pricing sheets, and modify labor units. Provides extensive labor unit tables and blank forms for your next electrical job.
272 pages, 8½ x 11, $19.00

National Construction Estimator

Current building costs for residential, commercial, and industrial construction. Estimated prices for every common building material. Manhours, recommended crew, and labor cost for installation. Includes an electronic version of the book on computer disk with a stand-alone *Windows* estimating program **FREE** on a 3½" high-density (1.44 Mb) disk. If you need 5¼" high-density disks add $10 extra.
528 pages, 8½ x 11, $37.50. Revised annually

Craftsman's Illustrated Dictionary of Construction Terms

Almost everything you could possibly want to know about any word or technique in construction. Hundreds of up-to-date construction terms, materials, drawings and pictures with detailed, illustrated articles describing equipment and methods. Terms and techniques are explained or illustrated in vivid detail. Use this valuable reference to check spelling, find clear, concise definitions of construction terms used on plans and construction documents, or learn about little-known tools, equipment, tests and methods used in the building industry. It's all here.
416 pages, 8½ x 11, $36.00

CD Estimator

If your computer has *Windows*™ and a CD-ROM drive, CD Estimator puts at your fingertips 80,000 construction costs for new construction, remodeling, electrical, plumbing, HVAC and painting. You'll also have the *National Estimator* program — a stand-alone estimating program for *Windows* that *Remodeling* magazine called a "computer wiz." Included is a communications program you can use to download cost updates on Contractor's Bulletin Board. To help you create professional-looking estimates, the disk includes over 40 construction estimating and bidding forms in a format that's perfect for nearly any word processing or spreadsheet program for *Windows.* And to top it off, a 70-minute interactive video teaches you how to use this CD-ROM to estimate construction costs. **CD Estimator is $59.00**

Profits in Buying & Renovating Homes

Step-by-step instructions for selecting, repairing, improving, and selling highly profitable "fixer-uppers." Shows which price ranges offer the highest profit-to-investment ratios, which neighborhoods offer the best return, practical directions for repairs, and tips on dealing with buyers, sellers, and real estate agents. Shows you how to determine your profit before you buy, what "bargains" to avoid, and how to make simple, profitable, inexpensive upgrades. **304 pages, 8½ x 11, $19.75**

Estimating Tables for Home Building

Produce accurate estimates for nearly any residence in just minutes. This handy manual has tables you need to find the quantity of materials and labor for most residential construction. Includes overhead and profit, how to develop unit costs for labor and materials, and how to be sure you've considered every cost in the job. **336 pages, 8½ x 11, $21.50**

Profits in Building Spec Homes

If you've ever wanted to make big profits in building spec homes yet were held back by the risks involved, you should have this book. Here you'll learn how to do a market study and feasibility analysis to make sure your finished home will sell quickly, and for a good profit. You'll find tips that can save you thousands in negotiating for land, learn how to impress bankers and get the financing package you want, how to nail down cost estimating, schedule realistically, work effectively yet harmoniously with subcontractors so they'll come back for your next home, and finally, what to look for in the agent you choose to sell your finished home. Includes forms, checklists, worksheets, and step-by-step instructions.
208 pages, 8½ x 11, $27.25

National Repair & Remodeling Estimator

The complete pricing guide for dwelling reconstruction costs. Reliable, specific data you can apply on every repair and remodeling job. Up-to-date material costs and labor figures based on thousands of jobs across the country. Provides recommended crew sizes; average production rates; exact material, equipment, and labor costs; a total unit cost and a total price including overhead and profit. Separate listings for high- and low-volume builders, so prices shown are specific for any size business. Estimating tips specific to repair and remodeling work to make your bids complete, realistic, and profitable. Includes an electronic version of the book on computer disk with a stand-alone *Windows* estimating program **FREE** on a 3½" high-density (1.44 Mb) disk. If you need 5¼" high-density disks add $10 extra.
416 pages, 11 x 8½, $38.50. Revised annually

NO POSTAGE
NECESSARY
IF MAILED
IN THE
UNITED STATES

BUSINESS REPLY MAIL
FIRST CLASS MAIL PERMIT NO. 271 CARLSBAD, CA

POSTAGE WILL BE PAID BY ADDRESSEE

Craftsman Book Company
6058 Corte del Cedro
P.O. Box 6500
Carlsbad, CA 92018-9974

NO POSTAGE
NECESSARY
IF MAILED
IN THE
UNITED STATES

BUSINESS REPLY MAIL
FIRST CLASS MAIL PERMIT NO. 271 CARLSBAD, CA

POSTAGE WILL BE PAID BY ADDRESSEE

Craftsman Book Company
6058 Corte del Cedro
P.O. Box 6500
Carlsbad, CA 92018-9974

NO POSTAGE
NECESSARY
IF MAILED
IN THE
UNITED STATES

BUSINESS REPLY MAIL
FIRST CLASS MAIL PERMIT NO. 271 CARLSBAD, CA

POSTAGE WILL BE PAID BY ADDRESSEE

Craftsman Book Company
6058 Corte del Cedro
P.O. Box 6500
Carlsbad, CA 92018-9974